Lecture Notes in Mathematics

Volume 2292

This series reports on new developments in all areas of mathematics and their applications - quickly, informally and at a high level. Mathematical texts analysing new developments in modelling and numerical simulation are welcome. The type of material considered for publication includes:

1. Research monographs
2. Lectures on a new field or presentations of a new angle in a classical field
3. Summer schools and intensive courses on topics of current research.

Texts which are out of print but still in demand may also be considered if they fall within these categories. The timeliness of a manuscript is sometimes more important than its form, which may be preliminary or tentative.

Titles from this series are indexed by Scopus, Web of Science, Mathematical Reviews, and zbMATH.

More information about this series at http://www.springer.com/series/304

Frank Neumann • Ambrus Pál

Editors

Homotopy Theory and Arithmetic Geometry – Motivic and Diophantine Aspects

LMS-CMI Research School, London, July 2018

 Springer

Editors
Frank Neumann
School of Computing and Mathematical
Sciences
University of Leicester
Leicester, UK

Ambrus Pál
Department of Mathematics
Imperial College London
London, UK

ISSN 0075-8434   ISSN 1617-9692   (electronic)
Lecture Notes in Mathematics
ISBN 978-3-030-78976-3   ISBN 978-3-030-78977-0   (eBook)
https://doi.org/10.1007/978-3-030-78977-0

Mathematics Subject Classification: 11G99, 14C17, 14C25, 14C35, 14F20, 14F35, 14F42, 14G05, 19E15, 19F99, 55M25, 55N99, 55P99, 55Q99

This Springer imprint is published by the registered company Springer Nature Switzerland AG.
The registered company address is: Gewerbestrasse 11, 6330 Cham, Switzerland

Preface

This present volume is based on lecture notes of the LMS-CMI Research School on *Homotopy Theory and Arithmetic Geometry: Motivic and Diophantine Aspects* and the Imperial College Nelder Fellow Lecture Series. Both activities were hosted at Imperial College in London, United Kingdom. The LMS-CMI research school took place 9–13 July 2018 and the Nelder Fellow Lecture Series prior to the school during 11–22 June 2018.

The focus of the LMS-CMI research school was on three major advances that have emerged recently in the interface between homotopy theory and arithmetic geometry: cohomological methods in intersection theory, with emphasis on motivic sheaves; homotopical obstruction theory for rational points and zero cycles; and arithmetic curve counts using motivic homotopy theory. The emergence of homotopical methods in arithmetic geometry represents one of the most important and exciting current trends in number theory, and these lecture notes aim to provide a gentle introduction to and overview on this fascinating and broad circle of ideas.

The main lecture series of the LMS-CMI research school were the following:

- Denis-Charles Cisinski (University of Regensburg, Germany): *Cohomological methods in intersection theory*
- Tomer Schlank (The Hebrew University of Jerusalem, Israel): *Homotopical manifestations of rational points and algebraic cycles*
- Kirsten Wickelgren (Duke University, USA): *Arithmetic enrichments of curve counts*

In addition to these lectures, the LMS-CMI research school featured four additional research talks:

- Jesse Kass (University of South Carolina, USA): *How to arithmetically count lines on a singular cubic surface*
- Paul Årne Østvær (University of Oslo, Norway): $\mathbb{A}^1$-*contractible varieties*
- Jon Pridham (University of Edinburgh, UK): *Iterative obstruction theory*
- Vesna Stojanoska (UIUC, USA): *A homotopical enhancement of Poitou-Tate duality*

Furthermore, as part of the Imperial College Nelder Fellow Lecture Series, the Nelder Visiting Fellow Paul Årne Østvær (University of Oslo, Norway) gave three research lectures on *Motivic Invariants* prior to the LMS-CMI research school, which featured a survey and a body of new results from motivic homotopy theory related to algebraic cobordism, motivic cohomology and motivic spheres. Starting with an overview on the foundations of motivic spaces and followed by a discussion on Milnor's influential conjectures relating K-theory to Galois cohomology and quadratic forms, the lectures focused on the constructions and calculations of universal motivic invariants.

The LMS-CMI Research Schools aim to provide training for young researchers in core modern research areas of mathematics. Doctoral students and postdoctoral researchers have the possibility to meet a number of world leading experts and other young researchers working in related areas. The LMS-CMI Research School at Imperial College featured over 40 participants from all over the world. It was run in partnership with the London Mathematical Society (LMS) and the Clay Mathematics Institute (CMI) with additional support from the Heilbronn Institute for Mathematical Research (HIMR) and Imperial College London. The organisers would like to thank all these institutions and organisations for their financial and logistic support. A big thank you goes to the authors for their important contributions and the fantastic job of producing these excellent notes, as well as to the expert referees for their fine work and additional suggestions. Moreover, we would like to thank all lecturers, speakers and participants who made this LMS-CMI research school and the Nelder Fellow Lecture Series such a great success. Finally, we like to thank the editors of Springer Lecture Notes in Mathematics for accepting this volume to be included in this prestigious series. Last but not least, many thanks go to Remi Lodh who guided us smoothly with expertise and patience through the editing and publishing process.

London, UK   Frank Neumann
May 2021   Ambrus Pál

Contents

Chapter 1
Homotopy Theory and Arithmetic Geometry—Motivic and Diophantine Aspects: An Introduction

Frank Neumann and Ambrus Pál

Abstract We give a brief overview on the various themes and topics of the lecture notes assembled in this volume and summarise each individual contribution based on the abstracts and introductions.

1.1 Overview of Themes

Applying methods of homotopy theory is one of the most important trends in arithmetic geometry in recent years. Motivic or $\mathbb{A}^1$-homotopy theory [18, 48, 51] was largely invented by Voevodsky to be used as a tool in order to prove the celebrated Milnor and Bloch-Kato conjectures [11, 47] in motivic cohomology [66–68]. Motivic cohomology and in particular Chow groups of smooth schemes are represented by Voevodsky's motivic Eilenberg-MacLane spectrum which is an object in the motivic stable homotopy category, similarly as classical singular cohomology in algebraic topology is represented by the Eilenberg-MacLane spectrum in the classical stable homotopy category. As in classical homotopy theory, motivic spectra represent other so-called generalised motivic cohomology theories like algebraic cobordism, algebraic K-theory or motivic stable homotopy [42, 43, 66]. Computations with motivic stable homotopy groups provide a lot of interesting arithmetic information but are even more challenging than those notoriously difficult calculations with their classical algebraic topological counterparts as can be seen for example from recent deep results by Röndigs, Spitzweck, and Østvær [58]. In this fashion one can also embark on studying algebraic varieties from a motivic homotopy point of view, for example studying $\mathbb{A}^1$-connected, $\mathbb{A}^1$-equivalent or $\mathbb{A}^1$-

F. Neumann (✉)
School of Computing and Mathematical Sciences, University of Leicester, Leicester, UK
e-mail: fn8@leicester.ac.uk; fn8@le.ac.uk

A. Pál
Department of Mathematics, Imperial College London, London, UK
e-mail: a.pal@imperial.ac.uk

© The Author(s), under exclusive license to Springer Nature Switzerland AG 2021
F. Neumann, A. Pál (eds.), *Homotopy Theory and Arithmetic Geometry – Motivic and Diophantine Aspects*, Lecture Notes in Mathematics 2292,
https://doi.org/10.1007/978-3-030-78977-0_1

1

contractible varieties, which also relates to birational geometry [3, 4, 17]. Blending homotopy theory and algebraic geometry has recently allowed for a variety of interesting applications in number theory and provides many future directions in attacking arithmetic questions via homotopy theory. For example, work of Hopkins and Wickelgren on the vanishing of Massey products [30] suggests that this technology can be used to solve a whole variety of embedding problems in Galois theory. Motivic homotopy theory has also had a major impact on arithmetic geometry in other directions, for example via the conjectural properties of a still undefined motivic fundamental group, which plays an integral part in Kim's formulation of a non-abelian Chabauty method to study rational points on curves [37]. In other groundbreaking work, Ayoub [6, 7] and Cisinski and Déglise [15, 16] generalised Voevodsky's definition of motives over a field to more general bases and extended the formalism of the six operations of Grothendieck to these objects, getting for example a very general Grothendieck-Verdier duality for constructible motivic sheaves with rational coefficients. This provides new insight into ℓ-adic realisations of motives and motivic interpretations of intersection theory and algebraic cycles.

Motivic homotopy theory is also playing a prominent role in extending classical enumerative geometry and intersection theory over the complex numbers to other fields which are of fundamental importance in number theory as can be seen for example in recent work by Asok, Fasel, Hoyois, Kass, Levine, Wendt, and Wickelgren [5, 22, 31, 33–35, 39, 41, 69]. Enumerative geometry is the classical branch of algebraic geometry concerned with counting numbers of solutions to geometric questions, mainly by means of intersection theory and its algebraic invariants the Chow groups. For example, famous classical enumerative problems include the classical count of 27 lines on a smooth complex cubic surfaces by Salmon and Cayley or Steiner's conic problem of the count of 3264 smooth conics tangent to five given conics in general position in the complex projective plane [20]. Instead of an integer count in the classical setting, counting in motivically enriched enumerative geometry produces a characteristic class in the Grothendieck-Witt ring of quadratic forms over a field, whose dimension then recovers the integer given by the classical count [10, 39, 40]. This encompasses similar counting phenomena in real algebraic geometry where signs appear when working over real closed fields corresponding to signatures of quadratic forms [23, 52]. Many constructions from algebraic and differential topology, like the degree [19, 46], have extensions in this new framework and can be utilised in this motivic enrichment of enumerative geometry [33, 57]. Particular useful and computable motivic invariants for motivic enumerative geometry are given by the Chow-Witt groups, which assemble a generalised motivic cohomology theory sitting in between motivic cohomology and stable motivic homotopy [22]. This motivic view towards enumerative geometry does not only provide arithmetic enrichments of classical counts but also gives exciting new arithmetic perspectives on other enumerative questions motivated by mathematical physics, symplectic geometry and tropical geometry like Gromov-Witten theory or motivic Donaldson-Thomas invariants [38].

Homotopy theory itself has also seen something of a revolution in recent years, with ground-breaking work by Lurie, Toën-Vezzosi and others on developing homotopical and derived algebraic geometry and so providing a whole new setting in which to study and unify problems in homotopy theory and algebraic geometry by using techniques of both [27, 44, 45, 62–64]. These advances are also intimately connected with questions of a more number theoretic flavour; one of the most striking applications of Lurie's foundational work is his proof with Gaitsgory of the Tamagawa number conjecture over function fields [26]. Another appearance of homotopy theory in arithmetic geometry is Grothendieck's famous section conjecture, which states that rational points on hyperbolic curves are controlled by their fundamental group. Since then homotopical techniques and invariants have proven to be a rich source of inspiration in Diophantine geometry, a good recent example being the work of Harpaz-Schlank [29] and Pál [55, 56] on applications of étale homotopy to the existence of rational points and the section conjecture, in particular in reinterpreting known obstructions to the local-global principle in purely homotopy-theoretical terms. Étale homotopy theory for algebraic varieties was originally developed by Artin and Mazur [2] and later refined by Friedlander [25] for simplicial schemes. Other recent applications and extensions of étale homotopy theory concern the homotopy types of algebraic stacks and higher stacks, including moduli stacks of polarised abelian varieties [13, 14, 24], as well as generalisations of anabelian geometry manifested for example in the groundbreaking work of Schmidt and Stix [59].

This collection of lecture notes from the LMS-CMI research school and the Nelder Fellow Lecture Series focuses on four of these major advances and core themes that have emerged recently in the interface between homotopy theory and arithmetic geometry: arithmetic enumerative geometry via motivic homotopy theory; cohomological methods in intersection theory with emphasis on motivic sheaves; homotopical obstruction theory for rational points and 0-cycles via étale homotopy theory and the study of algebraic varieties which are contractible in the sense of motivic homotopy theory. The emergence of homotopical methods in number theory represents one of the most important and exciting trends in contemporary mathematics and the many workshops and conferences organised recently around these topics demonstrate their timeliness and importance. We believe that the notes assembled here give a gentle and thorough introduction to these new trends and will inspire researchers and postgraduate students alike. The wide range and scope of these notes also indicate the broad intradisciplinary nature of the LMS-CMI research school and the Nelder Fellow Lecture Series around its core themes and document many exciting new connections and directions.

1.2   Summaries of Individual Contributions

As a quick orientation for the reader, we will give brief overviews on the topics and main themes of the individual contributions in this volume based on the abstracts and introductions.

Tom Brazelton: An Introduction to $\mathbb{A}^1$-Enumerative Geometry (Based on the LMS-CMI research school lectures by Kirsten Wickelgren)

These lecture notes provide an expository introduction to $\mathbb{A}^1$-enumerative geometry, which uses the machinery of $\mathbb{A}^1$-homotopy theory to extend classical enumerative geometry questions over the complex numbers to a broader range of fields and so giving rise to interesting counting problems in arithmetic geometry. The first part of this exposition starts with recollections on the classical topological degree from differential topology for morphisms of manifolds. Pursuing constructions originally due to Barge and Lannes, the topological notion of degree will then be enhanced to produce a notion of a degree valued in the Grothendieck-Witt ring $GW(k)$ of any field k. In order to produce such a degree more generally for morphisms of smooths schemes some machinery from $\mathbb{A}^1$-homotopy is then developed with the necessary details. A brief detour is provided to establish the general setting in which one can study motivic spaces, define in analogy to algebraic topology the unstable motivic homotopy category and make some basic but important calculations with colimits of motivic spaces. This part culminates with a discussion of the purity theorem of Morel and Voevodsky, which is needed to construct the local $\mathbb{A}^1$-degree due to Morel. The second part gives an exposition on the definition of this local $\mathbb{A}^1$-degree, allowing to enrich problems from classical enumerative geometry over the complex numbers to arbitrary fields, therefore giving rise to motivic enumerative geometry. A main part is then devoted to survey several important recent results and developments in $\mathbb{A}^1$-enumerative geometry. A detailed discussion is given on the Eisenbud-Khimshiashvili-Levine signature formula [21, 36] and its relation with the $\mathbb{A}^1$-degree as shown by Kass–Wickelgren [34]. Furthermore, enriched versions of the classical Milnor numbers in $\mathbb{A}^1$-homotopy theory due to Kass–Wickelgren [33] are discussed. The notes end with some recent computational results in $\mathbb{A}^1$-enumerative geometry by Kass–Wickelgren [35] on the number of lines on cubic surfaces over arbitrary fields extending the classical count of 27 lines on a complex cubic surface and by Srinivasan–Wickelgren [61] on a similar arithmetic count of the number of lines which meet four lines in 3-space.

Denis-Charles Cisinski: Cohomological Methods in Intersection Theory

Grothendieck and his school developed ℓ-adic cohomology to prove the Weil conjectures. In particular, for each prime number ℓ, one can apply cohomological methods to ℓ-adic cohomology in order to define invariants of schemes, such as Euler-Poincaré characteristics or ζ-functions. This means that we interpret algebraic cycles in terms of ℓ-adic cohomology classes. A main aspect of these lectures is to explain how motivic sheaves, as introduced by Voevodsky [65], allow to understand these invariants in a more intrinsic way: in the motivic world, there is no difference between algebraic cycles and cohomology classes. It will be briefly outlined what a motive is and indicated how the theory of motives can be used to enhance cohomological methods in intersection theory and so giving rise to natural ways for proving independence of ℓ-adic results and constructions of characteristic classes as 0-cycles. This leads naturally to the Grothendieck-Lefschetz formula, of which a new motivic proof is given. Furthermore, a proof of Grothendieck-Verdier duality for étale motives on schemes of finite type over a regular quasi-excellent scheme is given extending the level of generality in the existing literature. Another aspect of these notes is related to the study of $\mathbb{Q}$-linear motivic sheaves. For example it is proven that these motivic sheaves are in fact virtually integral. Another part is devoted to a proof for the motivic generic base change formula. The theory of motives used in these notes are h-motives over a scheme X, which are one of the many incarnations of étale motives. These motives are the objects of a triangulated category $DM_h(X)$ constructed and studied in detail in [15]. They are a natural non-effective modification of an earlier construction by Voevodsky [65] following the general theory of stable $\mathbb{A}^1$-homotopy theory of schemes as constructed by Morel and Voevodsky [51]. On the way it will also be shown that étale motives with torsion coefficients may be identified with classical étale sheaves. In particular, when restricted to the case of torsion coefficients, the results on trace formulas as discussed in these notes go essentially back to Grothendieck [28]. The case of rational coefficients has also been studied previously to some extend by Olsson [53, 54]. All these different cases fit naturally together as statements about étale motives with arbitrary coefficient systems. Finally, the motivic Grothendieck-Lefschetz formula is deduced from the Lefschetz-Verdier trace formula by using Bondarko's theory of weights [12] and Olsson's local term computations with the motivic Lefschetz-Verdier trace [53, 54].

Tomer M. Schlank: Étale Homotopy and Obstructions to Rational Points

The classical obstructions to the existence of rational points of algebraic varieties over number fields can be united and reinterpreted by using étale homotopy theory. These kind of obstructions can in fact be applied to sections of maps of arbitrary topoi. Furthermore this general framework can be used to understand closely related notions such as new obstructions to the existence of 0-cycles, to solutions to certain embedding problems in Galois theory and to other interesting problems in arithmetic geometry. These lecture notes give a condensed but approachable guide towards applications of étale homotopy theory to the study of rational points on algebraic varieties. They are aimed at readers with some background in algebraic geometry, but who are not necessarily familiar with modern homotopical algebra and ∞-categorical methods. Étale homotopy was originally introduced by Artin and Mazur [2] and further developed by Friedlander [25]. In recent years there has been a lot of activity around étale homotopy and its applications, in particular for applications in arithmetic geometry [1, 60]. In order to apply the machinery of étale homotopy to study homotopy obstructions for the existence of rational points Harpaz and Schlank [29] developed a relative version of étale homotopy. This relative version of étale homotopy requires a good notion of sheaves of homotopy types and a good model structure on pro-categories of simplicial objects in topoi [8, 9]. This notion is highly non-trivial as maps of homotopy types are on the one hand defined only up to homotopy, but on the other hand, for the obstruction theory to work, these homotopies need to be chosen in a certain coherent way. The correct notion to be used turns out to be that of an ∞-topos [45], which is itself a special case of an ∞-category. The notes start by giving an introduction and overview of the theory of ∞-categories followed by a discussion of ∞-topoi and their associated relative homotopy types. The next chapter features a review of the classical obstruction theory for sections of maps of homotopy types. This is followed by a discussion of the theory of étale topoi which produces ∞-topoi from schemes giving rise to the associated étale homotopy types. The final part of these notes features some explicit applications of the general theory to rational points and 0-cycles on varieties as well as to embedding problems in Galois theory.

Aravind Asok, Paul Årne Østvær: $\mathbb{A}^1$-homotopy theory and contractible varieties: a survey

This survey is partially based on the Nelder Fellow Lecture Series by the second author and a survey on some recent developments in $\mathbb{A}^1$-homotopy theory. The main goal is to highlight the interplay between $\mathbb{A}^1$-homotopy theory and affine algebraic geometry, focusing on those algebraic varieties that are contractible from various points of view. In particular, a main emphasis is given on the theory of algebraic varieties that are weakly contractible from the standpoint of $\mathbb{A}^1$-homotopy theory as developed by Morel and Voevodsky [18, 48–51, 66]. The first chapter gives an overview of results and questions from the geometry of contractible algebraic varieties, which follows a quick review of some aspects and results of the theory of open contractible manifolds from classical geometric topology as a way of inspiration for the algebro-geometric framework. The second chapter features a user's guide to $\mathbb{A}^1$-homotopy theory, in particular a construction of the unstable $\mathbb{A}^1$-homotopy category as well as a snapshot of the stable $\mathbb{A}^1$-homotopy category. The third chapter features fundamental examples of isomorphisms in the unstable $\mathbb{A}^1$-homotopy category and an overview of some fundamental results from affine and quasi-affine algebraic geometry in the context of motivic homotopy theory. In the fourth chapter further explicit calculations are presented involving $\mathbb{A}^1$-connectedness and rationality questions [4]. In the last chapter cancellation questions and $\mathbb{A}^1$-contractibility are discussed including some aspects of the theory of complex algebraic varieties whose associated complex manifolds are contractible [17, 32]. Along the way of this survey several related open problems are formulated and briefly discussed.

References

1. E. Arad, S. Carmeli, T.M. Schlank, Étale homotopy obstructions of arithmetic spheres, preprint (2019). arXiv:1902.03404
2. M. Artin, B. Mazur, Étale homotopy, in *Lecture Notes in Mathematics*, vol. 100 (Springer, Berlin, 1969)

3. A. Asok, F. Morel, Smooth varieties up to $\mathbb{A}^1$-homotopy and algebraic h-cobordisms. Adv. Math. **227**, 1990–2058 (2011)
4. A. Asok, Birational invariants and $\mathbb{A}^1$-connectedness. J. Reine Angew. Math. **681**, 39–64 (2013)
5. A. Asok, J. Fasel, Comparing Euler classes. Q. J. Math. **67**, 603–606 (2016)
6. J. Ayoub, Les six opérations de Grothendieck et le formalisme des cycles évanescents dans le monde motivique, I. Astérisque **314**, x+466 (2007)
7. J. Ayoub, Les six opérations de Grothendieck et le formalisme des cycles évanescents dans le monde motivique, II. Astérisque **315**, vi+364 (2007)
8. I. Barnea, T.M. Schlank, A projective model structure on pro-simplicial sheaves, and the relative étale homotopy type. Adv. Math. **291**, 784–858 (2016)
9. I. Barnea, Y. Harpaz, G. Horel, Pro-categories in homotopy theory. Algebr. Geom. Topol **17**, 567–643 (2017)
10. F. Binda, M. Levine, M.T. Nguyen, O. Röndigs (eds.), *Motivic Homotopy Theory and Refined Enumerative Geometry*. Workshop, Universität Duisburg-Essen, Essen, Germany, May 14–18, Contemporary Mathematics, vol. 745 (American Mathematical Society (AMS), Providence, 2020)
11. S. Bloch, K. Kato, *p*-adic étale cohomology. Publ. Math. Inst. Hautes Études Sci. **63**, 107–152 (1986)
12. M.V. Bondarko, Weights for relative motives: relation with mixed complexes of sheaves. Int. Math. Res. Not. IMRN **17**, 4715–4767 (2014)
13. D. Carchedi, On the étale homotopy type of higher stacks, preprint (2015). arXiv:1511.07830
14. C.-Y. Chough, Topological Types of Algebraic Stacks. Int. Math. Res. Not. **10**, 7799–7849 (2021)
15. D.-C. Cisinski, F. Déglise, Étales motives. Compositio Math. **152**, 556–666 (2016)
16. D.-C. Cisinski, F. Déglise, Triangulated categories of mixed motives, in *Springer Monographs in Mathematics* (Springer, Berlin, 2019), xlii+406 pp.
17. A. Dubouloz, S. Pauli, P.A. Østvær, $\mathbb{A}^1$-contractibility of affine modifications. Internat. J. Math. **30**, 1950069 (2019)
18. B.I. Dundas, M. Levine, P.A. Østvær, O. Röndigs, V. Voevodsky, Motivic homotopy theory, in *Lectures at a Summer School in Nordfjordeid, Norway, August 2002* (Universitext, Springer, New York, 2007)
19. D. Eisenbud, An algebraic approach to the topological degree of a smooth map. Bull. Am. Math. Soc. **84**, 751–764 (1978)
20. D. Eisenbud, J. Harris, *3264 and All That—A Second Course in Algebraic Geometry* (Cambridge University Press, Cambridge 2016)
21. D. Eisenbud, H.I. Levine, An algebraic formula for the degree of a C^∞-map germ. Ann. Math. **106**, 19–38 (1977)
22. J. Fasel, Groupes de Chow-Witt. Mem. Soc. Math. Fr. (N.S.) **113**, viii+197pp. (2008)
23. S. Finashin, V. Kharlamov, Abundance of real lines on real projective hypersurfaces, Int. Math. Res. Not. IMRN **16**, 3639–3646 (2013)
24. P. Frediani, F. Neumann, Étale homotopy types of moduli stacks of polarised abelian schemes. J. Homotopy Rel. Struct. **11**, 775–801 (2016)
25. E.M. Friedlander, Étale homotopy of simplicial schemes, in *Annals of Mathematics Studies*, vol. 104 (Princeton University, Princeton, 1982)
26. D. Gaitsgory, J. Lurie, *Weil's Conjecture for Function Fields*, vol. I. Annals of Mathematics Studies, vol. 199 (Princeton University, Princeton, 2019)
27. J.P.C. Greenlees, ed., *Axiomatic, enriched and motivic homotopy theory*. NATO Science Series II Mathematical Physics Chemistry, vol. 131 (Kluwer Academic Publication, Dordrecht, 2004)
28. A. Grothendieck, *Cohomologie l-adique et fonctions L*, *Séminaire de Géométrie Algébrique du Bois–Marie 1965–66 (SGA 5)*. Lecture Notes in Mathematics, vol. 589 (Springer, Berlin, 1977)
29. Y. Harpaz, T.M. Schlank, Homotopy obstructions to rational points, in *Torsors, Étale Homotopy and Applications to Rational Points*, ed. by A.N. Skorobogatov. London Mathematical Society Lecture Note Series, vol. 405 (Cambridge University, Cambridge, 2013), pp. 280–413

30. M.J. Hopkins, K. Wickelgren, Splitting varieties for triple Massey products. J. Pure Appl. Algebra **219**, 1304–1319 (2015)
31. M. Hoyois, A quadratic refinement of the Grothendieck-Lefschetz-Verdier trace formula. Algebr. Geom. Topol. **14**, 3603–3658 (2014)
32. M. Hoyois, A. Krishna, P.A. Østvær, $\mathbb{A}^1$-contractibility of Koras-Russell threefolds. Algebr. Geom. **3**, 407–423 (2016)
33. J. Kass, K. Wickelgren. $\mathbb{A}^1$-*Milnor Number, Oberwolfach Report*, vol. 35 (2016)
34. J. Kass, K. Wickelgren, The class of Eisenbud-Khimshiashvili-Levine is the local $\mathbb{A}^1$-Brouwer degree. Duke Math. J. **168**, 429–469 (2019)
35. J. Kass, K. Wickelgren, An arithmetic count of the lines on a smooth cubic surface. Compositio Math. **157**(4), 677–709 (2021). arXiv:1708.01175
36. G.N. Khimshiashvili, The local degree of a smooth mapping. Sakharth. SSR Mecn. Akad. Moambe **85**, 309–312 (1977)
37. M. Kim, The motivic fundamental group of $\mathbb{P}^1 \setminus \{0, 1, \infty\}$ and the theorem of Siegel. Invent. Math. **161**, 629–656 (2005)
38. M. Kontsevich, Y. Soibelman, Motivic Donaldson-Thomas invariants: summary of results. *Mirror Symmetry and Tropical Geometry. Manhattan, KS, USA, December 13–17, 2008*, ed. by R. Castaño-Bernard et al. (American Mathematical Society (AMS), Providence, 2010). Contemporary Mathematics **527**, 55–89 (2010)
39. M. Levine, Motivic Euler characteristics and Witt-valued characteristic classes. Nagoya Math. J. **236**, 251–310 (2019)
40. M. Levine, Lectures on quadratic enumerative geometry, in *Motivic Homotopy Theory and Refined Enumerative Geometry*, ed. by F. Binda, M. Levine, M.T. Nguyen, O. Röndigs. Workshop, Universität Duisburg-Essen, Essen, Germany, May 14–18, 2018 (American Mathematical Society (AMS), Providence, 2020). Contemporary Mathematics **745**, 163–198 (2020)
41. M. Levine, Aspects of enumerative geometry with quadratic forms. Doc. Math. **25**, 2179–2239 (2020)
42. M. Levine, F. Morel, Cobordisme algébrique I and II. C. R. Acad. Sci. Paris Sér. I Math. **332**, 723–728; 815–820 (2001)
43. M. Levine, F. Morel, *Algebraic Cobordism* (Springer, Berlin, 2007)
44. J. Lurie, *Derived Algebraic Geometry*, Ph.D. thesis (Massachusetts Institute of Technology, Cambridge, 2004), 193 pp. http://hdl.handle.net/1721.1/30144
45. J. Lurie, Higher topos theory, in *Annals of Mathematics Studies*, vol. 170 (Princeton University, Princeton, 2009)
46. J.W. Milnor, Topology from the differentiable viewpoint, in *Based on Notes*, ed. by D.W. Weaver (The University Press of Virginia, Charlottesville, 1965)
47. J.W. Milnor, Algebraic K-theory and quadratic forms. Inventiones Math. **9**, 318–344 (1970)
48. F. Morel, An introduction to $\mathbb{A}^1$-homotopy theory, in *Contemporary Developments in Algebraic K-theory*, ed. by M. Karoubi, A.O. Kuku, C. Pedrini. ICTP Lectures Notes, vol. XV (Abdus Salam International Centre for Theoretical Physics (ICTP), Trieste, 2004), pp. 357–441
49. F. Morel, $\mathbb{A}^1$-algebraic topology, in *Proceedings of the International Congress of Mathematicians*, vol. II, Zürich (2006), pp. 1035–1059
50. F. Morel, $\mathbb{A}^1$-algebraic topology over a field, in *Lecture Notes in Mathematics*, vol. 2052 (Springer, Berlin, 2012)
51. F. Morel, V. Voevodsky, $\mathbb{A}^1$-homotopy theory of schemes. Publ. Math. Inst. Hautes Études Sci. **90**, 45–143 (2001)
52. C. Okonek, A. Teleman, Intrinsic signs and lower bounds in real algebraic geometry. J. Reine Angew. Math. **688**, 219–241 (2014)
53. M. Olsson, Borel-Moore homology, Riemann-Roch transformations, and local terms. Adv. Math. **273**, 56–123 (2015)
54. M. Olsson, Motivic cohomology, localized Chern classes, and local terms. Manuscripta Math. **149**, 1–43 (2016)

55. A. Pál, The real section conjecture and Smith's fixed point theorem for pro-spaces. J. Lond. Math. Soc. **83**, 353–367 (2011)
56. A. Pál, Étale homotopy equivalence of rational points on algebraic varieties. Algebra Number Theory **9**, 815–873 (2015)
57. S. Pauli, K. Wickelgren, Applications to $\mathbb{A}^1$-enumerative geometry of the $\mathbb{A}^1$-degree. Res. Math. Sci. **8**(2), 1–21 (2021)
58. O. Röndigs, M. Spitzweck, P.A. Østvær, The first stable homotopy groups of motivic spheres. Ann. Math. **189**, 1–74 (2019)
59. A. Schmidt, J. Stix, Anabelian geometry with étale homotopy types. Ann. Math. **184**, 817–868 (2016)
60. A.N. Skorobogatov (ed.), Torsors, Étale Homotopy and applications to rational points, in *London Mathematical Society Lecture Note Series*, vol. 405 (Cambridge University, Cambridge, 2013)
61. P. Srinivasan, K. Wickelgren (with an appendix by B. Kadets, P. Srinivasan, A.A. Swaminathan, L. Taylor, and D. Tseng), An arithmetic count of the lines meeting four lines in $\mathbb{P}^3$. Trans. AMS (to appear). arXiv:1810.03503
62. B. Toën, G. Vezzosi, From HAG to DAG: derived moduli stacks, in *Axiomatic, Enriched and Motivic Homotopy Theory*, ed. by J.P.C. Greenlees. NATO Science Series II Mathematical Physics Chemistry, vol. 131 (Kluwer Acadamic Publication, Dordrecht, 2004), pp. 173–216
63. B. Toën, G. Vezzosi, Homotopical algebraic geometry i: topos theory. Adv. Math. **193**, 257–372 (2005)
64. B. Toën, G. Vezzosi, Homotopical algebraic geometry ii: geometric stacks and applications. Mem. Am. Math. Soc. **193**(902), x+224 pp. (2008)
65. V. Voevodsky, Homology of schemes. Selecta Math. **2**, 111–153 (1996)
66. V. Voevodsky, $\mathbb{A}^1$-homotopy theory, in *Proceedings of the International Congress of Mathematicians*, vol. I (Springer, Berlin 1998), pp. 579–604
67. V. Voevodsky, Reduced power operations in motivic cohomology. Publ. Math. Inst. Hautes Études Sci. **98**, 1–57 (2003)
68. V. Voevodsky, Motivic cohomology with $\mathbb{Z}/2$-coefficients. Publ. Math. Inst. Hautes Études Sci. **98**, 59–104 (2003)
69. M. Wendt, Oriented Schubert calculus in Chow–Witt rings of Grassmannians, in *Motivic Homotopy Theory and Refined Enumerative Geometry. Workshop, Universität Duisburg-Essen, Essen, Germany, May 14–18, 2018*, ed. by F. Binda, M. Levine, M.T. Nguyen, O. Röndigs. Contemporary Mathematics, vol. 745 (American Mathematical Society (AMS), Providence, 2020), pp. 217–267

Chapter 2
An Introduction to $\mathbb{A}^1$-Enumerative Geometry

Based on Lectures by Kirsten Wickelgren Delivered at the LMS-CMI Research School "Homotopy Theory and Arithmetic Geometry—Motivic and Diophantine Aspects"

Thomas Brazelton

Abstract We provide an expository introduction to $\mathbb{A}^1$-enumerative geometry, which uses the machinery of $\mathbb{A}^1$-homotopy theory to enrich classical enumerative geometry questions over a broader range of fields. Included is a discussion of enriched local degrees of morphisms of smooth schemes, following Morel, $\mathbb{A}^1$-Milnor numbers, as well as various computational tools and recent examples.

2.1 Introduction

In the late 1990s Fabien Morel and Vladimir Voevodsky investigated the question of whether techniques from algebraic topology, particularly homotopy theory, could be applied to study varieties and schemes, using the affine line $\mathbb{A}^1$ rather than the interval [0, 1] as a parametrizing object. This idea was influenced by a number of preceding papers, including work of Karoubi and Villamayor [1] and Weibel [2] on K-theory, and Jardine's work on algebraic homotopy theory [3, 4]. In work with Suslin developing an algebraic analog of singular cohomology which was $\mathbb{A}^1$-invariant [5], Voevodsky laid out what he considered to be the starting point of a homotopy theory of schemes parametrized by the affine line [6]. This relied upon Quillen's theory of model categories [7], which provided the abstract framework needed to develop homotopy theory in broader contexts. Following this work, Morel [8] and Voevodsky [6] constructed equivalent unstable $\mathbb{A}^1$-homotopy categories, laying the groundwork for their seminal paper [9] which marked the

T. Brazelton (✉)
University of Pennsylvania, David Rittenhouse Lab., Philadelphia, PA, USA
e-mail: tbraz@math.upenn.edu

© The Author(s), under exclusive license to Springer Nature Switzerland AG 2021   11
F. Neumann, A. Pál (eds.), *Homotopy Theory and Arithmetic Geometry – Motivic and Diophantine Aspects*, Lecture Notes in Mathematics 2292,
https://doi.org/10.1007/978-3-030-78977-0_2

genesis of $\mathbb{A}^1$-*homotopy theory*. Since its inception, this field of mathematics has seen far-reaching applications, perhaps most notably Voevodsky's resolution of the Bloch-Kato conjecture, a classical problem from number theory [10].

The machinery of $\mathbb{A}^1$-homotopy theory works over an arbitrary field k (in fact over arbitrary schemes, and even richer mathematical objects), allowing enrichments of classical problems which have only been explored over the real and complex numbers. Recent work in this area has generalized classical enumerative problems over wider ranges of fields, forming a body of work which we are referring to as $\mathbb{A}^1$-*enumerative geometry*.

Beginning with a recollection of the topological degree for a morphism between manifolds in Sect. 2.2.1, we pursue an idea of Barge and Lannes to produce a notion of degree valued in the Grothendieck–Witt ring of a field k, defined in Sect. 2.2.2. We produce such a naive degree for endomorphisms of the projective line in Sect. 2.2.3, however in order to produce such a degree for smooth n-schemes in general, we will need to develop some machinery from $\mathbb{A}^1$-homotopy theory. A brief detour is taken to establish the setting in which one can study motivic spaces, defining the unstable motivic homotopy category in Sect. 2.2.4, and establishing some basic, albeit important computations involving colimits of motivic spaces in Sect. 2.2.5. This discussion culminates in the purity theorem of Morel and Voevodsky, stated in Sect. 2.2.6, which is requisite background for defining the local $\mathbb{A}^1$-degree following Morel.

In Sect. 2.3, we are finally able to define the local $\mathbb{A}^1$-degree of a morphism of schemes, which is a powerful, versatile tool in enriching enumerative geometry problems over arbitrary fields. At this point, we survey a number of recent results in $\mathbb{A}^1$-enumerative geometry. We discuss the Eisenbud–Khimshiashvili–Levine signature formula in Sects. 2.3.1 and 2.3.2, and we see its relation to the $\mathbb{A}^1$-degree, as proved in [11]. An enriched version of the $\mathbb{A}^1$-Milnor number is provided in Sect. 2.3.3, which provides an enriched count of nodes on a hypersurface, following [12]. The problem of counting lines on a cubic surface, and the associated enriched results [13] are discussed in Sect. 2.3.4. Finally, in Sect. 2.3.5 we provide an arithmetic count of lines meeting four lines in three-space, following [14].

Throughout these conference proceedings, various exercises (most of which were provided by Wickelgren in her 2018 lectures) are included. Detailed solutions may be found on the author's website.

2.2 Preliminaries

2.2.1 Enriching the Topological Degree

A continuous map $f : S^n \to S^n$ from the n-sphere to itself induces a homomorphism on the top homology group $f_* : H_n(S^n) \to H_n(S^n)$, which is of the form $f_*(x) = dx$ for some $d \in \mathbb{Z}$. This integer d defines the *(global) degree* of the map

f. If f and g are homotopic as maps from the n-sphere to itself, they will induce the same homomorphism on homology groups. Therefore, taking $[S^n, S^n]$ to denote the set of homotopy classes of maps, we can consider degree as a function

$$\mathrm{deg}^{\mathrm{top}} : [S^n, S^n] \to \mathbb{Z}.$$

Throughout these notes, we will use the notation $\mathrm{deg}^{\mathrm{top}}$ to refer to the topological (integer-valued) degree.

For any continuous map of n-manifolds $f : M \to N$, we could define a naive notion of the "local degree" around a point $p \in M$ via the following procedure: suppose that $q \in N$ has the property that $f^{-1}(q)$ is discrete, and let $p \in f^{-1}(q)$. Pick a small ball W containing q, and a small ball $V \subseteq f^{-1}(W)$ satisfying $V \cap f^{-1}(q) = \{p\}$. Then we may see that the spaces $V/(V \smallsetminus \{p\}) \simeq (V/\partial V) \simeq S^n$ are homotopy equivalent. Similarly, we have that $W/(W \smallsetminus \{q\}) \simeq S^n$. We obtain the following diagram:

$$
\begin{array}{ccc}
S^n & \xrightarrow{\ \ g\ \ } & S^n \\
{\scriptstyle\simeq}\downarrow & & \uparrow{\scriptstyle\simeq} \\
V/(V \smallsetminus \{p\}) & \xrightarrow{\ \ f\ \ } & W/(W \smallsetminus \{q\}).
\end{array}
\tag{2.1}
$$

Thus we could define the local (topological) degree of f around our point p, denoted $\mathrm{deg}_p^{\mathrm{top}}(f)$, to be the induced degree map on the n-spheres, that is, $\mathrm{deg}_p^{\mathrm{top}}(f) := \mathrm{deg}^{\mathrm{top}}(g)$ in the diagram above. If $f^{-1}(q) = \{p_1, \ldots, p_m\}$ is a discrete set of isolated points, we may relate the global degree to the local degree via the following formula

$$\mathrm{deg}^{\mathrm{top}}(f) = \sum_{i=1}^{m} \mathrm{deg}_{p_i}^{\mathrm{top}}(f).$$

One may prove that the left hand side is independent of q, and thus that the choice of q is arbitrary in calculating the global degree from local degrees. In differential topology, when discussing the degree of a locally differentiable map f between n-manifolds, we have a simple formula for the local degree at a simple zero. We pick local coordinates $(x_1, \ldots, x_n)$ in a neighborhood of our point p_i, and local coordinates $(y_1, \ldots, y_n)$ around a regular value q. Then we can interpret f locally as a map $f = (f_1, \ldots, f_n) : \mathbb{R}^n \to \mathbb{R}^n$. Suppose that the Jacobian $\mathrm{Jac}(f)$ is nonvanishing at the point p_i. Then we define

$$\mathrm{deg}_{p_i}^{\mathrm{top}}(f) = \mathrm{sgn}(\mathrm{Jac}(f)(p_i)) = \begin{cases} +1 & \text{if } \mathrm{Jac}(f)(q_i) > 0 \\ -1 & \text{if } \mathrm{Jac}(f)(q_i) < 0. \end{cases}$$

When working over a field k, Barge and Lannes[1] defined a notion of degree for a map $\mathbb{P}_k^1 \to \mathbb{P}_k^1$. Their insight was, rather than taking the sign of the Jacobian as in differential topology, to instead remember the value of $\mathrm{Jac}(f)(p_i)$ as a square class in $k^\times / (k^\times)^2$. Over the reals this recovers the sign, but over a general field we may have vastly more square classes. We encode this value as a rank one symmetric bilinear form over k, and we will soon see that this idea can be used to define a local degree at k-rational points, and that by using field traces we can extend the definition of local degree to hold for points with residue fields a finite separable extension of k. These degrees, rather than being integers, are elements of the *Grothendieck–Witt ring of k*, denoted $\mathrm{GW}(k)$, defined below.

2.2.2 The Grothendieck–Witt Ring

Over a field k, we may form a semiring of isomorphism classes of non-degenerate symmetric bilinear forms (or quadratic forms if we assume $\mathrm{char}(k) \neq 2$) on vector spaces over k, using the operations $\otimes_k$ and $\oplus$. Group completing this semiring with respect to $\oplus$, we obtain the Grothendieck–Witt ring $\mathrm{GW}(k)$. For any $a \in k^\times$, we let $\langle a \rangle \in \mathrm{GW}(k)$ denote the following rank one bilinear form:

$$\langle a \rangle : k \times k \to k$$

$$(x, y) \mapsto axy.$$

Symmetric bilinear forms are equivalent if they differ only by a change of basis. For example, if $b \neq 0$ we can see that $\langle ab^2 \rangle (x, y) = \langle a \rangle (bx, by)$, so we identify $\langle a \rangle = \langle ab^2 \rangle$ in $\mathrm{GW}(k)$, since these bilinear forms agree up to a vector space automorphism of k. We may describe $\mathrm{GW}(k)$ to be a ring generated by elements $\langle a \rangle$ for each $a \in k^\times / (k^\times)^2$, subject to the following relations [16, Lemma 4.9]

1. $\langle a \rangle \langle b \rangle = \langle ab \rangle$
2. $\langle a \rangle + \langle b \rangle = \langle ab(a + b) \rangle + \langle a + b \rangle$, for $a + b \neq 0$
3. $\langle a \rangle + \langle -a \rangle = \langle 1 \rangle + \langle -1 \rangle$. We conventionally denote this element as $\mathbb{H} := \langle 1 \rangle + \langle -1 \rangle$, called the *hyperbolic element* of $\mathrm{GW}(k)$.

Exercise 1 In the statements above, (1) and (2) imply (3).

Proposition 1 *We have a ring isomorphism $\mathrm{GW}(\mathbb{C}) \cong \mathbb{Z}$, given by taking the rank.*

Proof We remark that $\langle a \rangle = \langle b \rangle$ for any $a, b \in \mathbb{C}^\times$, thus we only have one isomorphism class of non-degenerate symmetric bilinear forms in rank one. □

---

[1] Unpublished. See the note by Morel on [15, p.1037].

The isomorphism $GW(\mathbb{C}) \cong \mathbb{Z}$ relates to a general fact that the $\mathbb{A}^1$-degree of a morphism of complex schemes recovers the size of the fiber, counted with multiplicity.

Proposition 2 *The rank and signature provide a group isomorphism $GW(\mathbb{R}) \cong \mathbb{Z} \times \mathbb{Z}$.*

Proof The Gram matrix of a symmetric bilinear form on $\mathbb{R}^n$ is an $n \times n$ real symmetric matrix A. After diagonalizing our matrix A, we can always find a change of basis in which the eigenvalues lie in the set $\{-1, 0, 1\}$. A non-degenerate symmetric bilinear form guarantees that no eigenvalues will vanish, so all of these eigenvalues will be ± 1. We may define the *signature* of A as the number of 1's appearing on the diagonalized matrix minus the number of -1's, and by *Sylvester's law of inertia* this determines an invariant on our matrix A. Thus we obtain an injective map

$$GW(\mathbb{R}) \to \mathbb{Z} \times \mathbb{Z}$$

$$A \mapsto (\text{rank}(A), \text{sig}(A)).$$

The image of this map is the subgroup $\{(a + b, a - b) : a, b \in \mathbb{Z}\}$, which one may verify is isomorphic to $\mathbb{Z} \times \mathbb{Z}$.   □

The multiplication on $GW(\mathbb{R})$ does not agree with that of $\mathbb{Z} \times \mathbb{Z}$, in the sense that $GW(\mathbb{R}) \cong \mathbb{Z} \times \mathbb{Z}$ is not a ring isomorphism. However one may verify that the map

$$GW(\mathbb{R}) \to \frac{\mathbb{Z}[t]}{(t^2 - 1)},$$

given by sending $\langle 1 \rangle \mapsto 1$ and $\langle -1 \rangle \mapsto t$, is in fact a ring isomorphism, and hence provides the multiplicative structure of $GW(\mathbb{R})$.

Proposition 3 *The rank and determinant provide a group isomorphism $GW(\mathbb{F}_q) \cong \mathbb{Z} \times \mathbb{F}_q^\times / (\mathbb{F}_q^\times)^2$.*

Proof sketch We may still use the rank of our matrix as an invariant for $GW(\mathbb{F}_q)$. Additionally, we might use the determinant of our matrix to distinguish between symmetric bilinear forms. However note that, for any similar matrix $C^T A C$, it has determinant $\det(C^T A C) = \det(A) \det(C)^2$. Therefore, we can view the determinant as a well-defined map $\det : GW(\mathbb{F}_q) \to \mathbb{F}_q^\times / (\mathbb{F}_q^\times)^2$. After group completion, we obtain a map $GW(\mathbb{F}_q) \xrightarrow{(\text{rank,det})} \mathbb{Z} \times \mathbb{F}_q^\times / (\mathbb{F}_q^\times)^2$, which we verify is a group isomorphism. For more details, see [17, II,Theorem 3.5].   □

One may use $GW(\mathbb{F}_q)$ to understand $GW(\mathbb{Q}_p)$ by applying the following result.

Theorem 1 ([17, VI,Theorem 1.4] Springer's Theorem) *Let K be a complete discretely valued field, and κ be its residue field, with the assumption that char(κ) $\neq$ 2. Then there is an isomorphism of groups*

$$GW(K) \cong \frac{GW(\kappa) \oplus GW(\kappa)}{\mathbb{Z}[\mathbb{H}, -\mathbb{H}]}.$$

Corollary 1 *We have a group isomorphism* $GW(\mathbb{C}((t))) = \mathbb{Z} \oplus \mathbb{Z}/2$.

We should see how the Grothendieck–Witt ring interacts with extensions of fields. For a separable field extension $K \subset L$, and an element $\beta \in GW(L)$, we can view the composition

$$V \times V \xrightarrow{\beta} L \xrightarrow{\mathrm{Tr}_{L/K}} K$$

as an element of $GW(K)$ by post-composing with the trace map $L \to K$, and considering V as a K-vector space. This provides us a natural homomorphism between Grothendieck–Witt rings[2]

$$\mathrm{Tr}_{L/K} : GW(L) \to GW(K).$$

At the level of $\mathbb{A}^1$-homotopy theory, this trace comes from a transfer on stable homotopy groups—for more detail see [16, §4]. Now that we have seen some computations involving the Grothendieck–Witt ring, we can develop in detail the notion of degree for maps of schemes.

2.2.3 Lannes' Formula

Let $f : \mathbb{P}_k^1 \to \mathbb{P}_k^1$ be a non-constant endomorphism of the projective line over a field of characteristic 0. We can then pick a rational point $q \in \mathbb{P}_k^1$, with fiber $f^{-1}(q) = \{p_1, \ldots, p_m\}$ such that $\mathrm{Jac}(f)(p_i) \neq 0$ for each i, where the Jacobian is computed by picking the same affine coordinates on both copies of $\mathbb{P}_k^1$. Since $\mathrm{Jac}(f)(p_i) \in k(p_i)$ is only defined in a residue field, we must precompose with the trace map in order to define the *local $\mathbb{A}^1$-degree*

$$\deg_{p_i}^{\mathbb{A}^1} f := \mathrm{Tr}_{k(p_i)/k} \langle \mathrm{Jac}(f)(p_i) \rangle. \tag{2.2}$$

We can then define the global $\mathbb{A}^1$-degree of f as the following sum, which is independent of our choice of rational point q with discrete fiber (this fact is

---

[2]When the field extension is assumed to be finite but the separability condition is dropped, a more general notion of transfer is given by *Scharlau's transfer* [17, VII §1].

attributable to Lannes and Morel, although a detailed proof may be found in [11, Proposition 14]):

$$\deg^{\mathbb{A}^1} f := \sum_{i=1}^{m} \mathrm{Tr}_{k(p_i)/k} \langle \mathrm{Jac}(f)(p_i) \rangle.$$

Exercise 2   Compute the $\mathbb{A}^1$-degrees of the following maps:

1. $\mathbb{P}_k^1 \to \mathbb{P}_k^1$, given by $z \mapsto az$, for $a \in k^\times$.
2. $\mathbb{P}_k^1 \to \mathbb{P}_k^1$, given by $z \mapsto z^2$.

Maps of schemes $\mathbb{P}_k^1 \to \mathbb{P}_k^1$ are precisely rational functions $\frac{f}{g}$. Assuming that f and g are relatively prime, we can determine the classical topological (integer-valued) degree of this rational function as

$$\deg^{\mathrm{top}}\left(\frac{f}{g}\right) = \max\{\deg^{\mathrm{top}}(f), \deg^{\mathrm{top}}(g)\}.$$

To the rational function $\frac{f}{g}$, one may associated a bilinear form, called the *Bézout form*, which is denoted $\mathrm{Béz}\left(\frac{f}{g}\right)$. This is done by introducing two variables X and Y, and remarking that we have the following equality

$$\frac{f(X)g(Y) - f(Y)g(X)}{X - Y} = \sum_{1 \le i, j \le n} B_{ij} X^{i-1} Y^{j-1},$$

where $n = \deg^{\mathrm{top}}\left(\frac{f}{g}\right)$, and where $B_{ij} \in k$. We can see that this defines a symmetric bilinear form $k^n \times k^n \to k$, whose Gram matrix is given by the coefficients B_{ij}.

Exercise 3   Compute the Bézout bilinear forms of the maps given in Exercise 2.

Theorem 2 (Cazanave)   *We have that*

$$Béz\left(\frac{f}{g}\right) = \deg^{\mathbb{A}^1}\left(\frac{f}{g}\right).$$

This is stated in [18, Theorem 2], but is attributable to [19].

This provides us with an efficient way to compute the $\mathbb{A}^1$-degree of rational maps while circumventing the tedium of computing the local $\mathbb{A}^1$-degree at each point in a fiber.

2.2.4 The Unstable Motivic Homotopy Category

One of the primary ideas in $\mathbb{A}^1$-homotopy theory is to replace the unit interval in classical homotopy theory with the affine line $\mathbb{A}^1_k = \mathrm{Spec}\,(k[t])$. To this end, one might develop a *naive $\mathbb{A}^1$-homotopy* of maps of schemes $f, g : X \to Y$ as a morphism

$$h : X \times \mathbb{A}^1_k \to Y,$$

such that $h(x, 0) = f(x)$ and $h(x, 1) = g(x)$ for all $x \in X$. This was first introduced by Karoubi and Villamayor [1]. This notion of naive $\mathbb{A}^1$-homotopy is not generally the most effective, partially due to the following observation.

Exercise 4 [20] Prove that naive $\mathbb{A}^1$-homotopy fails to be a transitive relation on hom-sets by considering three morphisms $\mathrm{Spec}\,k \to \mathrm{Spec}\,k[x, y]/(xy)$ identifying the points $(0, 1)$, $(0, 0)$, and $(1, 0)$.

We will build a model category in which we have a class of $\mathbb{A}^1$-*weak equivalences*, and we will denote by $[-, -]_{\mathbb{A}^1}$ the weak equivalence classes of morphisms. In particular, naive $\mathbb{A}^1$-homotopy equivalences are tractable examples of $\mathbb{A}^1$-weak equivalences. Nonetheless, naive $\mathbb{A}^1$-homotopy generates an equivalence relation, and in practice the naive homotopy classes of maps $[X, Y]_N$ are often easier to compute than their genuine counterparts $[X, Y]_{\mathbb{A}^1}$. In fact, the naive homotopy classes of maps $[\mathbb{P}^1_k, \mathbb{P}^1_k]_N$ are equipped with an addition, induced by pinch maps, which endows this set with a monoid structure. It was demonstrated by Cazanave that the map

$$[\mathbb{P}^1_k, \mathbb{P}^1_k]_N \to [\mathbb{P}^1_k, \mathbb{P}^1_k]_{\mathbb{A}^1}$$

is a group completion [19].

In order to study the homotopy theory of schemes, we must develop a model structure which encodes a notion of $\mathbb{A}^1$-weak equivalence. In particular we must force $\mathbb{A}^1$ to be contractible—as we have remarked, the initial motivation for forming such a model category was to treat $\mathbb{A}^1$ as if it were akin to the interval $[0, 1]$ in the category of topological spaces. Morel and Voevodsky initially formulated the theory of the "homotopy category of a site with an interval"; for this classical treatment see [9, §2.3].

We remark that the category of smooth k-schemes Sm_k does not admit all colimits, and therefore cannot be endowed with a model structure. To rectify this issue, we pass to the category of the simplicial presheaves via the Yoneda embedding

$$Sm_k \to \mathrm{sPre}(Sm_k) = \mathrm{Fun}(Sm_k^{\mathrm{op}}, sSet)$$

$$X \mapsto \mathrm{Map}(-, X).$$

This new category is *cocomplete* (it admits all small colimits), and moreover can be equipped with the *projective model structure* arising from the classical model structure on simplicial sets. Given our model structure, we are now permitted to identify a class of morphisms which we would like to call weak equivalences, and perform Bousfield localization in order to formally invert them. For exposition on Bousfield localization and related results, we refer the reader to [21].

The analog of open covers in a categorical setting is provided by a Grothendieck topology τ. The category Sm_k can be equipped with a Grothendieck topology in order to make it a site, after which, we will apply Bousfield localization to render the class of τ-hypercovers (our analog of open covers) into weak equivalences. We remark that by Dugger et al. [22, Theorem 6.2], this localization exists, and we denote it by $L_\tau : \mathrm{sPre}(Sm_k) \to \mathrm{Sh}_{\tau,k}$. The fibrant objects in $\mathrm{Sh}_{\tau,k}$ are those simplicial presheaves which are *homotopy sheaves* in the τ topology [23, p. 20]. We therefore think about the localization L_τ as a way to encode the topology τ into the homotopy theory of $\mathrm{sPre}(Sm_k)$.

Due to the wealth of properties granted to us by simplicial presheaves, the category $\mathrm{Sh}_{\tau,k}$ inherits a left proper combinatorial simplicial model category structure, and in particular we are allowed to perform Bousfield localization again in order to force $\mathbb{A}^1$ to be contractible. We identify a set of maps $\{X \times \mathbb{A}^1 \to X\}$, indexed over the set of isomorphism classes of objects in Sm_k, as our desired weak equivalences, then perform a final Bousfield localization $L_{\mathbb{A}^1}$ with respect to this set. Finally, we define

$$\mathrm{Spc}_{\tau,k}^{\mathbb{A}^1} := L_{\mathbb{A}^1}\mathrm{Sh}_{\tau,k} = L_{\mathbb{A}^1}L_\tau\mathrm{sPre}(Sm_k).$$

This category has a model structure by construction, and we refer to its homotopy category as the *unstable motivic homotopy category*. Throughout these notes and in much of the literature, it is assumed we are using the *Nisnevich topology* (which is defined and contrasted with other choices of topologies below), and we will write $\mathrm{Spc}_k^{\mathbb{A}^1} := \mathrm{Spc}_{\mathrm{Nis},k}^{\mathbb{A}^1}$. Our primary objects of study in $\mathrm{Spc}_k^{\mathbb{A}^1}$ will be the fibrant objects of this category, which we refer to as $\mathbb{A}^1$-*spaces*. These admit a tangible recognition as precisely those presheaves which are valued in Kan complexes, satisfy Nisnevich descent, and are $\mathbb{A}^1$-invariant [23, Remark 3.58]. For more detail, see [23, §3].

There are many equivalent constructions of $\mathrm{Spc}_k^{\mathbb{A}^1}$, one notable one arising from the *universal homotopy theory* on the category of smooth schemes, as described by Dugger et al. [24]. By freely adjoining homotopy colimits, we obtain a universal category $U(Sm_k)$ which we may localize at the collections of maps

$$\{\mathrm{hocolim}U_\bullet \to X \ : \ \{U_\alpha\} \text{ is a hypercover of } X\}$$

$$\{X \times \mathbb{A}^1 \to X\}.$$

This procedure produces a model category $U(Sm_k)_{\mathbb{A}^1}$ which is Quillen equivalent to $\mathrm{Spc}_k^{\mathbb{A}^1}$.

Remark 1 In more modern language, one may build $\mathrm{Spc}_k^{\mathbb{A}^1}$ using $(\infty, 1)$-categories rather than model categories. Such a perspective may be found throughout the literature, for example in [25, 26].

One may study the categories $\mathrm{Spc}_{k,\tau}^{\mathbb{A}^1}$ arising from other choices of Grothendieck topologies, and indeed the homotopy theories arising from each selection behave quite differently and merit individual study. A small inexhaustive list of possible topologies includes the Zariski, Nisnevich, and étale topologies.

Definition 1 Suppose that X and Y are smooth over a field k. Then we say $f : X \to Y$ is *étale* at x if the induced map on cotangent spaces

$$(f^* \Omega_{Y/k})_x \xrightarrow{\cong} \Omega_{X/k,x}$$

is an isomorphism [27, §2.2, Corollary 10]. If we have the additional structure of coordinates on our base and target spaces, this is equivalent to the condition that $\mathrm{Jac}(f) \neq 0$ in $k(x)$.

For example, any open immersion $X \hookrightarrow Y$ is a local isomorphism, and is therefore an étale map.

Definition 2 Let $\{f_\alpha : U_\alpha \to X\}$ be a family of étale morphisms. We say that it is

1. an *étale cover* if this is a cover of X, that is the underlying map of topological spaces is surjective
2. a *Nisnevich cover* if this is a cover of X, and for every $x \in X$ there exists an $\alpha \in A$ and $y \in U_\alpha$ such that $y \mapsto x$ and it induces an isomorphism on residue fields $k(y) \xrightarrow{\cong} k(x)$
3. a *Zariski cover* if this is a cover of X, and each f_α is an open immersion.

Remark 2 Every Zariski cover is a Nisnevich cover, and every Nisnevich cover is an étale cover, however the converses of these statements do not hold.

In the Nisnevich topology, we are also able to retain some of the advantages that the Zariski topology offers. One of the primary advantages is that algebraic K-theory satisfies Nisnevich descent. Additionally we are able to compute the Nisnevich cohomological dimension as the Krull dimension of a scheme [9, p.94]. Finally, we refer the reader to [23, Proposition 7.2], which allows us to treat morphisms of schemes locally as morphisms of affine spaces, analogous to charts of Euclidean space in differential topology.

2.2.5   Colimits

Recall that the primary motivation in passing from Sm_k to $\mathrm{sPre}(Sm_k)$ was the existence of colimits. Despite the fact that Sm_k does not admit all small colimits, it still admits some—as a class of examples, consider colimits of schemes arising from Zariski open covers. The problem is that the Yoneda embedding $y : Sm_k \to \mathrm{sPre}(Sm_k)$ does not preserve colimits in general, thus in our efforts to rectify the failure of Sm_k to admit colimits, we have essentially forgotten about the colimits that it did in fact possess. This is part of the motivation to localize at τ-hypercovers—we see that colimits of schemes correspond to hypercovers on the associated representable presheaves. By our discussion in the previous section, the localization L_τ can be considered as the localization precisely at the class of maps $\mathrm{hocolim} U_\bullet \to X$ for any τ-hypercover $U_\bullet \to X$. Thus colimits of schemes are recorded in the category $\mathrm{Spc}_k^{\mathbb{A}^1}$ as homotopy colimits corresponding to hypercovers. For ease of reference, we summarize this in the following slogan.

Slogan Colimits of smooth schemes along τ-covers yield homotopy colimits of motivic spaces.

To illustrate this point, we consider the following example, where $\mathbb{G}_m := \mathrm{Spec}\, k\left[x, \frac{1}{x}\right]$ denotes the multiplicative group scheme.

Example 1 Let $f : \mathbb{G}_m \to \mathbb{A}_k^1$ be given by $z \mapsto z$, and $g : \mathbb{G}_m \to \mathbb{A}_k^1$ be given by $z \mapsto \frac{1}{z}$. Then the diagram

$$
\begin{array}{ccc}
\mathbb{G}_m & \xrightarrow{\;f\;} & \mathbb{A}_k^1 \\
{\scriptstyle g}\downarrow & & \downarrow \\
\mathbb{A}_k^1 & \longrightarrow & \mathbb{P}_k^1
\end{array}
$$

is a homotopy pushout of motivic spaces.

Proof We see that the two copies of the affine line form a Zariski open cover of $\mathbb{P}_k^1$, and hence a Nisnevich open cover of schemes. This corresponds to a hypercover on the representable simplicial presheaves, and after localization at Nisnevich hypercovers, we see that the homotopy pushout of $\left(\mathbb{A}_k^1 \leftarrow \mathbb{G}_m \to \mathbb{A}_k^1\right)$ is precisely $\mathbb{P}_k^1$. $\square$

For based topological spaces, recall we have a smash product, defined as

$$
X \wedge Y = X \times Y \big/ \left((X \times \{y\}) \cup (\{x\} \times Y)\right).
$$

We can think about the category of based topological spaces as the slice category $*/\mathrm{Top}$, where $*$ denotes the one-point space, i.e. the terminal object. By similarly taking the slice category under the terminal object $* := \mathrm{Spec}\, k$, we obtain a pointed version of $\mathrm{Spc}_k^{\mathbb{A}^1}$, which is often denoted by $\mathrm{Spc}_{k,*}^{\mathbb{A}^1}$.[3] We can then define the smash product as the homotopy cofiber of the canonical map between the coproduct of two pointed motivic spaces into their product:

$$
\begin{array}{ccc}
X \vee Y & \longrightarrow & X \times Y \\
\downarrow & & \downarrow \\
* & \longrightarrow & X \wedge Y.
\end{array}
$$

One may define the *suspension* as $\Sigma X := S^1 \wedge X$, which we may verify is the same as the homotopy cofiber of $X \to *$. One may see that, since $\mathbb{A}_k^1 \simeq \mathrm{Spec}\, k$ is contractible, we have that Example 1 implies that $\mathbb{P}_k^1$ is the homotopy cofiber of the unique map $\mathbb{G}_m \to \mathrm{Spec}\, k$. Concisely, this example tells us that

$$
\mathbb{P}_k^1 \simeq \Sigma \mathbb{G}_m.
$$

Recall from topology that the spheres satisfy $S^n \wedge S^m \cong S^{n+m}$. In developing a homotopy theory of schemes, we would like to search for a class of objects satisfying an analogous property. From this motivation, we uncover two types of spheres in $\mathrm{Spc}_k^{\mathbb{A}^1}$. The first, denoted S^1, is called the *simplicial sphere*, and can be thought of as the union of three copies of the affine line, enclosing a triangle. As a simplicial presheaf, we think of it as the constant presheaf at $S^1 = \Delta^1/\partial\Delta^1$. Our second sphere, often called the *Tate sphere*, is taken to be the projective line $\mathbb{P}_k^1 \simeq S^1 \wedge \mathbb{G}_m$.

There are various conventions for the notation on spheres in $\mathbb{A}^1$-homotopy theory, and in the literature one may see $S^{p+q\alpha}$, $S^{p,q}$ or $S^{p+q,q}$ to mean the same thing, depending on the context. In these notes, we will use the convention that

$$
S^{p+q\alpha} := (S^1)^{\wedge p} \wedge (\mathbb{G}_m)^{\wedge q}.
$$

Exercise 5 Show that the diagram

$$
\begin{array}{ccc}
X \times Y & \longrightarrow & X \\
\downarrow & & \downarrow \\
Y & \longrightarrow & \Sigma(X \wedge Y)
\end{array}
$$

is a homotopy pushout diagram. The context for this example is left ambiguous as the result holds in $\mathrm{Spc}_{k,*}^{\mathbb{A}^1}$ just as well as it does for pointed topological spaces.

Example 2 There is an $\mathbb{A}^1$-homotopy equivalence $\mathbb{A}_k^n \setminus \{0\} \simeq (S^1)^{\wedge(n-1)} \wedge (\mathbb{G}_m)^{\wedge n}$.

---

[3]We note that such a slice category must be taken at the level of model categories rather than homotopy categories in order to have a tractable pointed homotopy theory.

Proof Note that we may construct $\mathbb{A}_k^n \smallsetminus \{0\}$ as a homotopy pushout

$$
\begin{array}{ccc}
(\mathbb{A}_k^1 \smallsetminus \{0\}) \times (\mathbb{A}_k^{n-1} \smallsetminus \{0\}) & \longrightarrow & \mathbb{A}_k^1 \times (\mathbb{A}_k^{n-1} \smallsetminus \{0\}) \\
\downarrow & & \downarrow \\
(\mathbb{A}_k^1 \smallsetminus \{0\}) \times \mathbb{A}_k^n & \longrightarrow & \mathbb{A}_k^n \smallsetminus \{0\} .
\end{array}
$$

Applying the exercise above, we see that

$$
\mathbb{A}_k^n \smallsetminus \{0\} \simeq \Sigma(\mathbb{A}_k^{n-1} \smallsetminus \{0\}) \wedge (\mathbb{A}_k^1 \smallsetminus \{0\}) = S^1 \wedge (\mathbb{A}_k^{n-1} \smallsetminus \{0\}) \wedge \mathbb{G}_m.
$$

The result follows inductively. $\square$

Notation For a morphism of motivic spaces $f : X \to Y$, denote by Y/X the homotopy cofiber of the map f, that is, the homotopy pushout

$$
\begin{array}{ccc}
X & \xrightarrow{\ f\ } & Y \\
\downarrow & & \downarrow \\
* & \longrightarrow & Y/X.
\end{array}
$$

Example 3 (Excision)   Suppose that X is a smooth scheme over k, that $Z \hookrightarrow X$ is a closed immersion, and that $U \supseteq Z$ is a Zariski open neighborhood of Z inside of X. Then we have a Nisnevich weak equivalence (that is, a weak equivalence in the category $\mathrm{Sh}_{\mathrm{Nis},k}$)

$$
\frac{U}{U \smallsetminus Z} \xrightarrow{\ \sim\ } \frac{X}{X \smallsetminus Z}.
$$

We refer to this result informally as *excision* (not to be confused with excision in the sense of [23, Proposition 3.53]), as we regard this weak equivalence as excising the closed subspace $X \smallsetminus U$ from the top and bottom of the cofiber $X/(X \smallsetminus Z)$.

Proof We remark that $(X \smallsetminus Z)$ and U form a Zariski open cover of X, and that their intersection is $(X \smallsetminus Z) \cap U = U \smallsetminus Z$. As Zariski covers are Nisnevich covers, one remarks that we have a homotopy pushout diagram of motivic spaces

$$
\begin{array}{ccc}
U \smallsetminus Z & \longrightarrow & X \smallsetminus Z \\
\downarrow & & \downarrow \\
U & \longrightarrow & X.
\end{array}
$$

The fact that the homotopy cofibers of the vertical maps in the diagram above are $\mathbb{A}^1$-weakly equivalent follows from the following diagram:

$$
\begin{array}{ccccc}
U \smallsetminus Z & \longrightarrow & X \smallsetminus Z & \longrightarrow & * \\
\downarrow & & \downarrow & & \downarrow \\
U & \longrightarrow & X & \longrightarrow & \dfrac{X}{X \smallsetminus Z}.
\end{array}
$$

As the left and right squares are homotopy cocartesian, it follows formally that the entire rectangle is homotopy cocartesian. □

Example 4 There is an $\mathbb{A}^1$-homotopy equivalence $\mathbb{P}_k^n / \mathbb{P}_k^{n-1} \simeq (S^1)^{\wedge n} \wedge (\mathbb{G}_m)^{\wedge n}$

Proof As $\mathbb{P}_k^n \smallsetminus \{0\}$ is the total space of $O(1)$ on $\mathbb{P}_k^{n-1}$, we have an $\mathbb{A}^1$-equivalence $\mathbb{P}_k^n \smallsetminus \{0\} \simeq \mathbb{P}_k^{n-1}$. Therefore, one sees $\mathbb{P}_k^n / \mathbb{P}_k^{n-1} \simeq \mathbb{P}_k^n / (\mathbb{P}_k^n - \{0\})$. Via excision, we are able to excise everything away from a standard affine chart, from which we may see that $\mathbb{P}_k^n / (\mathbb{P}_k^n - \{0\}) \simeq \mathbb{A}_k^n / (\mathbb{A}_k^n - \{0\})$. Contracting $\mathbb{A}_k^n$, we obtain $* / (\mathbb{A}_k^n - \{0\}) \simeq \Sigma(\mathbb{A}_k^n - \{0\})$. Therefore $\mathbb{P}_k^n / \mathbb{P}_k^{n-1} \simeq \Sigma(\mathbb{A}_k^n - \{0\}) \simeq (S^1)^{\wedge n} \wedge (\mathbb{G}_m)^{\wedge n}$ after applying Example 2. □

This last example is of particular interest, as it exhibits the cofiber $\mathbb{P}_k^n / \mathbb{P}_k^{n-1}$ as a type of sphere in $\mathbb{A}^1$-homotopy theory. Given an endomorphism of such a motivic sphere, Morel defined a *degree homomorphism*

$$
\deg^{\mathbb{A}^1} : \left[\mathbb{P}_k^n / \mathbb{P}_k^{n-1}, \mathbb{P}_k^n / \mathbb{P}_k^{n-1} \right]_{\mathbb{A}^1} \to \mathrm{GW}(k),
$$

which he proved was an isomorphism in degrees $n \geq 2$ [15, Corollary 4.11].

Recall that to define a local Brouwer degree of an endomorphism between n-manifolds, we first had to pick a ball containing a point p, and then identify the cofiber $W/(W \smallsetminus \{p\})$ with the n-sphere S^n. This allowed us to construct Diagram 2.1, after which we could apply the degree homomorphism $[S^n, S^n] \to \mathbb{Z}$ to define a local degree. An analogous procedure will be available to us in $\mathbb{A}^1$-homotopy theory if, for a Zariski open neighborhood U around a k-rational point x, we are able to associate a canonical $\mathbb{A}^1$-weak equivalence between $U/(U \smallsetminus \{x\})$ and $\mathbb{P}_k^n / \mathbb{P}_k^{n-1}$. Indeed this is possible via the theorem of *purity*.

2.2.6 Purity

One of the major techniques in $\mathbb{A}^1$-homotopy theory comes from the purity theorem. In manifold topology, the tubular neighborhood theorem allows us to define a diffeomorphism between a tubular neighborhood of a smooth immersion and an open neighborhood around its zero section in the normal bundle. In $\mathbb{A}^1$ homotopy theory, the Nisnevich topology isn't fine enough to define such a tubular neighborhood, however we can still get an analog of the tubular neighborhood theorem which will allow us to define, among other things, local $\mathbb{A}^1$-degrees of maps.

Definition 3 A *Thom space* of a vector bundle $V \to X$ is the cofiber

$$V / (V \smallsetminus X),$$

where $V \smallsetminus X$ denotes the vector bundle minus its zero section. In the literature, this may be denoted by

$$\mathrm{Thom}(V, X) = \mathrm{Th}(V) = X^V.$$

Remark 3 We may also describe the Thom space of a vector bundle via an $\mathbb{A}^1$-weak equivalence

$$\mathrm{Th}(V) \simeq \frac{\mathrm{Proj}\,(V \oplus O)}{\mathrm{Proj}\,(V)}.$$

Proof We have a map $V \to V \oplus O$ sending $v \mapsto (v, 1)$, and we may view this inside of projective space via the inclusion $V \oplus O \subseteq \mathrm{Proj}\,(V \oplus O)$. Via excision (Example 3), we have a Nisnevich weak equivalence

$$\frac{\mathrm{Proj}\,(V \oplus O)}{\mathrm{Proj}\,(V \oplus O) \smallsetminus 0} \simeq \frac{V}{V \smallsetminus 0},$$

where 0 denotes the image of the zero section. We remark that $\mathrm{Proj}\,(V \oplus O) \smallsetminus 0$ is the total space of $O(-1)$ on $\mathrm{Proj}\,(V)$, thus we have an $\mathbb{A}^1$-weak equivalence $\mathrm{Proj}\,(V \oplus O) \smallsetminus 0 \simeq \mathrm{Proj}\,(V)$. The result follows from observing $\frac{\mathrm{Proj}\,(V \oplus O)}{\mathrm{Proj}\,(V \oplus O) \smallsetminus 0} \simeq \frac{\mathrm{Proj}\,(V \oplus O)}{\mathrm{Proj}\,(V)}$. $\square$

Theorem 3 (Purity Theorem) *Let $Z \hookrightarrow X$ be a closed immersion in Sm_k. Then we have an $\mathbb{A}^1$-equivalence*

$$\frac{X}{X \smallsetminus Z} \simeq Th(N_Z X),$$

where $N_Z X \to Z$ denotes the normal bundle of Z in X.

Proof The proof uses the deformation to the normal bundle of Fulton and MacPherson [28]. Let f denote the composition of the maps

$$\mathrm{Bl}_{Z \times \{0\}}(X \times \mathbb{A}_k^1) \to X \times \mathbb{A}_k^1 \to \mathbb{A}_k^1.$$

We define $D_Z X$ to be the scheme $\mathrm{Bl}_{Z \times \{0\}}(X \times \mathbb{A}_k^1) \smallsetminus \mathrm{Bl}_{Z \times \{0\}}(X \times \{0\})$, and note that f restricts to a map $f\Big|_{D_Z X} : D_Z X \to \mathbb{A}_k^1$. We may compute the fiber of $f|_{D_Z X}$ over 0 as

$$
\begin{aligned}
f\Big|_{D_Z X}^{-1}(0) &= \mathrm{Proj}\left(N_{Z \times \{0\}}(X \times \mathbb{A}_k^1)\right) \smallsetminus \mathrm{Proj}\left(N_{Z \times \{0\}}(X \times \{0\})\right) \\
&= \mathrm{Proj}(N_Z X \oplus O) \smallsetminus \mathrm{Proj}(N_Z X) \\
&= N_Z X,
\end{aligned}
$$

and the fiber over 1 as $f\Big|_{D_Z X}^{-1}(1) = X$. Since $Z \times \mathbb{A}_k^1$ determines a closed subscheme in $D_Z X$, we have that the fiber over 0 is $Z \subseteq N_Z X$ and the fiber over 1 is $Z \subseteq X$. Thus we obtain morphisms of pairs

$$
\begin{aligned}
(Z, N_Z X) &\xrightarrow{i_0} (Z \times \mathbb{A}_k^1, D_Z X) \\
(Z, X) &\xrightarrow{i_1} (Z \times \mathbb{A}_k^1, D_Z X),
\end{aligned}
\tag{2.3}
$$

corresponding to the inclusions of the fibers over the points 0 and 1, respectively. To prove the purity theorem, it now suffices to show that the induced morphisms on cofibers are weak equivalences:

$$
\frac{N_Z X}{N_Z X \smallsetminus Z} \to \frac{D_Z X}{D_Z X \smallsetminus Z \times \mathbb{A}_k^1}
$$

$$
\frac{X}{X \smallsetminus Z} \to \frac{D_Z X}{D_Z X \smallsetminus Z \times \mathbb{A}_k^1}.
$$

Lemma 1 *[23, Lemma 7.3] Suppose that* P *is a property of smooth pairs of schemes such that the following properties hold:*

1. *If* (Z, X) *is a smooth pair of schemes and* $\{U_\alpha \to X\}_{\alpha \in A}$ *is a Zariski cover of* X *such that* P *holds for the pair*

$$
(Z \times_X U_{\alpha_1} \times_X \cdots \times_X U_{\alpha_n}, U_{\alpha_1} \times_X \cdots \times_X U_{\alpha_n})
$$

   for each $(\alpha_1, \ldots, \alpha_n)$, *then* P *holds for* (Z, X)
2. *If* $(Z, X) \to (Z, Y)$ *is a morphism of smooth pairs inducing an isomorphism on* Z *such that* $X \to Y$ *is étale, then* P *holds for* (Z, X) *if and only if* P *holds for* (Z, Y)
3. P *holds for the pair* $(Z, \mathbb{A}_k^n \times Z)$,

then P *holds for all smooth pairs.*

To conclude the proof of purity, we let **P** be the property on the pair (Z, X) that the morphisms in Eq. (2.3) induce homotopy pushout diagrams[4]

$$
\begin{array}{ccc}
Z & \longrightarrow & \dfrac{N_Z X}{N_Z X \smallsetminus Z} \\[1em]
\downarrow & \ulcorner \quad \downarrow & \\[1em]
Z \times \mathbb{A}^1_k & \longrightarrow & \dfrac{D_Z X}{D_Z X \smallsetminus Z \times \mathbb{A}^1_k}
\end{array}
\qquad
\begin{array}{ccc}
Z & \longrightarrow & \dfrac{X}{X \smallsetminus Z} \\[1em]
\downarrow & \ulcorner \quad \downarrow & \\[1em]
Z \times \mathbb{A}^1_k & \longrightarrow & \dfrac{D_Z X}{D_Z X \smallsetminus Z \times \mathbb{A}^1_k}.
\end{array}
$$

One may check that Lemma 1 holds for this property, and therefore since $Z \to Z \times \mathbb{A}^1_k$ is a weak equivalence, a homotopy pushout along this map is also a weak equivalence. Thus we obtain a sequence of $\mathbb{A}^1$-weak equivalences

$$
\frac{X}{X \smallsetminus Z} \xrightarrow{\sim} \frac{D_Z X}{D_Z X \smallsetminus Z \times \mathbb{A}^1_k} \xleftarrow{\sim} \frac{N_Z X}{N_Z X \smallsetminus Z} = \mathrm{Th}(N_Z X).
$$

$\square$

2.3 $\mathbb{A}^1$-enumerative Geometry

As discussed above, Morel exhibited the global degree of maps of motivic spheres as

$$
\deg^{\mathbb{A}^1} : [\mathbb{P}^n_k/\mathbb{P}^{n-1}_k, \mathbb{P}^n_k/\mathbb{P}^{n-1}_k]_{\mathbb{A}^1} \to \mathrm{GW}(k).
$$

Recall that, for a scheme X, we have functors to the category of topological spaces obtained by taking real and complex points, that is, $X \mapsto X(\mathbb{R})$ and $X \mapsto X(\mathbb{C})$. Morel's degree map satisfies a compatibility diagram with the degree maps we recognize from algebraic topology[5]

$$
\begin{array}{ccccc}
[S^n, S^n] & \xleftarrow{\mathbb{R}\text{-pts}} & [\mathbb{P}^n_{\mathbb{R}}/\mathbb{P}^{n-1}_{\mathbb{R}}, \mathbb{P}^n_{\mathbb{R}}/\mathbb{P}^{n-1}_{\mathbb{R}}]_{\mathbb{A}^1} & \xrightarrow{\mathbb{C}\text{-pts}} & [S^{2n}, S^{2n}] \\[0.5em]
{\scriptstyle \deg^{\mathrm{top}}} \downarrow & & \downarrow {\scriptstyle \deg^{\mathbb{A}^1}} & & \downarrow {\scriptstyle \deg^{\mathrm{top}}} \\[0.5em]
\mathbb{Z} & \xleftarrow{\quad \mathrm{sig} \quad} & \mathrm{GW}(\mathbb{R}) & \xrightarrow{\quad \mathrm{rank} \quad} & \mathbb{Z}.
\end{array}
\qquad (2.4)
$$

---

[4]Equivalently, one may say that i_0 and i_1 are *weakly excisive* morphisms of pairs [29, Definition 3.17].

[5]The commutativity of this diagram is one of the key features of Morel's $\mathbb{A}^1$-degree and is attributable to him [15, p. 1037]. We can provide an alternative justification of this fact following the discussion of the EKL form in Sect. 2.3.1.

We can apply the purity theorem to develop a notion of local degree for a general map between schemes of the same dimension. Suppose that $f : \mathbb{A}_k^n \to \mathbb{A}_k^n$, and $x \in \mathbb{A}_k^n$ is a k-rational preimage of a k-rational point $y = f(x)$. Further suppose that x is an isolated point in $f^{-1}(y)$, meaning that there exists a Zariski open set $U \subseteq \mathbb{A}_k^n$ such that $x \in U$ and $f^{-1}(y) \cap U = x$.

Definition 4 In the conditions above, the *local* $\mathbb{A}^1$-*degree* of f at x is defined to be the degree of the map

$$U \Big/ (U \smallsetminus \{x\}) \xrightarrow{\overline{f}} \mathbb{A}_k^n \Big/ (\mathbb{A}_k^n \smallsetminus \{y\}),$$

under the $\mathbb{A}^1$-weak equivalences $U \big/ (U \smallsetminus \{x\}) \cong \mathrm{Th}(T_x \mathbb{A}_k^n) \cong \mathbb{P}_k^n \big/ \mathbb{P}_k^{n-1}$ and $\mathbb{A}_k^n \big/ (\mathbb{A}_k^n \smallsetminus \{y\}) \cong \mathbb{P}_k^n \big/ \mathbb{P}_k^{n-1}$ provided to us by purity and by the canonical trivialization of the tangent space of affine space.

Dropping the assumption that $k(x) = k$, but still assuming that y is k-rational, we may equivalently define $\deg_x^{\mathbb{A}^1} f$ as the degree of the composite

$$\mathbb{P}_k^n \Big/ \mathbb{P}_k^{n-1} \to \mathbb{P}_k^n \Big/ (\mathbb{P}_k^n \smallsetminus \{x\}) \cong U \Big/ (U \smallsetminus \{x\}) \xrightarrow{\overline{f}} \mathbb{A}_k^n \Big/ (\mathbb{A}_k^n \smallsetminus \{y\}) \cong \mathbb{P}_k^n \Big/ \mathbb{P}_k^{n-1}.$$

Proposition 4 *These definitions of the local degree are equivalent. This was proven in [11, Prop. 12], which is a generalization of a proof of Hoyois [30, Lemma 5.5].*

Equation (2.2) admits the following generalization to endomorphisms of affine space.

Proposition 5 *[11, Proposition 15] Let $f : \mathbb{A}_k^n \to \mathbb{A}_k^n$, assume that f is étale at a closed point $x \in \mathbb{A}_k^n$, and assume that that $f(x) = y$ is k-rational and that x is isolated in its fiber. Then the local degree is given by*

$$\deg_x^{\mathbb{A}^1}(f) = Tr_{k(x)/k} \langle Jac(f)(x) \rangle.$$

Remark 4 At a non-rational point p whose residue field $k(p)|k$ is a finite separable extension of the ground field, the local $\mathbb{A}^1$-degree can be computed by base changing to $k(p)$ to compute the local degree rationally, and applying the field trace $Tr_{k(p)/k}$ to obtain a well-defined element of $GW(k)$ [31].

2.3.1 The Eisenbud–Khimshiashvili–Levine Signature Formula

Given a morphism $f = (f_1, \ldots, f_n) : \mathbb{A}_k^n \to \mathbb{A}_k^n$ with an isolated zero at the origin, we may associate to it a certain isomorphism class of bilinear forms $w_0^{\mathrm{EKL}}(f)$, called the *Eisenbud–Levine–Khimshiashvili (EKL) class*. This was studied by Eisenbud and Levine, and independently by Khimshiashvili, in the case where f is a smooth endomorphism of $\mathbb{R}^n$ [32, 33]. They ascertained that the degree $\deg_0^{\text{top}} f$ can be computed as the signature of the form $w_0^{\text{EKL}}(f)$. If f is furthermore assumed to be real analytic, the rank of this form recovers the degree of the complexification $f_{\mathbb{C}}$ [34]. This bilinear form $w_0^{\text{EKL}}(f)$ can be defined over an arbitrary field k, and in this setting Eisenbud asked the following question: does $w_0^{\text{EKL}}(f)$ have any topological interpretation? We will see that the answer is yes, via work of Kass and Wickelgren [11].

Suppose that $f = (f_1, \ldots, f_n) : \mathbb{A}_k^n \to \mathbb{A}_k^n$ has an isolated zero at the origin, and define the local k-algebra

$$Q_0(f) := \frac{k[x_1, \ldots, x_n]_{(x_1, \ldots, x_n)}}{(f_1, \ldots, f_n)}.$$

We may pick polynomials a_{ij} so that, for each i, we have the equality

$$f_i(x_1, \ldots, x_n) = f_i(0) + \sum_{j=1}^n a_{ij} \cdot x_j.$$

By taking their determinant, we define $E_0(f) := \det(a_{ij})$ as an element of $Q_0(f)$, which we refer to as the *distinguished socle element* of the local algebra $Q_0(f)$. We remark that when $\text{Jac}(f)$ is a nonzero element of $Q_0(f)$, one has the equality [35, 4.7 Korollar]

$$\text{Jac}(f) = \dim_k(Q_0(f)) \cdot E_0(f).$$

We then pick η to be any k-linear vector space homomorphism $\eta : Q_0(f) \to k$ satisfying $\eta(E_0(f)) = 1$. One may check that the following bilinear form

$$Q_0(f) \times Q_0(f) \to k$$
$$(u, v) \mapsto \eta(u \cdot v)$$

is non-degenerate and its isomorphism class is independent of the choice of η [32, Propositions 3.4,3.5] and [11, §3]. The class of this form in $\text{GW}(k)$ is referred to as the EKL class, and denoted by $w_0^{\text{EKL}}(f)$.

Example 5 If $f : \mathbb{A}_k^1 \to \mathbb{A}_k^1$ is given by $z \mapsto z^2$, we may see that $Q_0(f) = k[z]_{(z)}/(z^2)$. We see that f has an isolated zero at the origin, and that

$$f = f(0) + x \cdot x,$$

hence $E_0(f) = x$. We determine $\eta : Q_0(f) \to k$ on a basis for $Q_0(f)$ by setting $\eta(x) = 1$ and $\eta(1) = 0$. Then we compute the EKL form via its Gram matrix as:

$$\begin{pmatrix} \eta(1 \cdot 1) & \eta(1 \cdot x) \\ \eta(x \cdot 1) & \eta(x \cdot x) \end{pmatrix} = \begin{pmatrix} 0 & 1 \\ 1 & 0 \end{pmatrix} = \mathbb{H}.$$

Theorem 4 *If $f : \mathbb{A}_k^n \to \mathbb{A}_k^n$ is any endomorphism of affine space with an isolated zero at the origin, there is an equality* $\deg_0^{\mathbb{A}^1} f = w_0^{EKL}(f)$ *in GW(k) [11].*

In particular we observe that the compatibility stated in Diagram 2.4 is justified by this theorem, combined with the results of Eisenbud–Khimshiashvili–Levine and Palamodov. Moreover we remark that the EKL form can be defined at any k-rational point, and an analogous statement to Theorem 4 holds in this context.

Exercise 6

1. Compute the degree of $f : \mathbb{A}_k^2 \to \mathbb{A}_k^2$, given as $f(x, y) = (4x^3, 2y)$ in the case where char$(k) \neq 2$.
2. Supposing f is étale at the origin 0, show that $w_0^{EKL}(f) = \langle \mathrm{Jac}(f)(0) \rangle$ is an equality in GW(k). Show furthermore that an analogous equality holds at any k-rational point x.

As a generalization of Exercise 6(2), one may show that if f is étale at a point x, one has the following equality in GW(k)

$$w_x^{EKL}(f) = \mathrm{Tr}_{k(x)/k} \langle \mathrm{Jac}(f)(x) \rangle. \tag{2.5}$$

This is shown using Galois descent, as in [11, Lemma 33].

2.3.2 Sketch of Proof for Theorem 4

Step 1: We can see that $\deg_0^{\mathbb{A}^1} f$ and $w_0^{EKL}(f)$ are finitely determined in the sense that they are unchanged by changing f to $f + g$, with $g = (g_1, \ldots, g_n)$, and $g_i \in \mathfrak{m}_0^N$ for sufficiently large N, where $\mathfrak{m}_0 := (x_1, \ldots, x_n)$ denotes the maximal ideal at the origin [11, Lemma 17].

Step 2: By changing f to $f + g$, we may assume that f extends to a finite, flat morphism $F : \mathbb{P}_k^n \to \mathbb{P}_k^n$, where $F^{-1}(\mathbb{A}_k^n) \subseteq \mathbb{A}_k^n$ and $F|_{F^{-1}(0) \setminus \{0\}}$ is étale [11, Proposition 23].

Proposition 6 *(Scheja–Storch) [35, §3, pp.180–182] We have that $w_0^{EKL}(f)$ is a direct summand of the fiber at 0 of a family of bilinear forms over $\mathbb{A}_k^n$, which we construct below.*

We will prove Proposition 6 following the construction of this family of bilinear forms.

The Scheja–Storch Construction   Let $F : \mathrm{Spec}\,(P) \to \mathrm{Spec}\,(A)$, where

$$P = k[x_1, \ldots, x_n]$$
$$A = k[y_1, \ldots, y_n].$$

One may show that the collection $\{t_1, \ldots, t_n\}$ is a regular sequence in $A[x_1, \ldots, x_n]$, where $t_i := y_i - F_i(x_1, \ldots, x_n)$. Then

$$B = A[x_1, \ldots, x_n]\big/\langle t_1, \ldots, t_n \rangle$$

is a relative complete intersection, which parametrizes the fibers of F. This regular sequence determines a canonical isomorphism [35, Satz 3.3]

$$\theta : \mathrm{Hom}_A(B, A) \overset{\cong}{\to} B,$$

via the following procedure: we may first express

$$t_j \otimes 1 - 1 \otimes t_j = \sum_{i=1}^n a_{ij}\,(x_i \otimes 1 - 1 \otimes x_i),$$

where each a_{ij} is an element of $A[x_1, \ldots, x_n] \otimes_A A[x_1, \ldots, x_n]$. Under the projection map $A[x_1, \ldots, x_n] \otimes_A A[x_1, \ldots, x_n] \to B \otimes_A B$, we have that $\det\,(a_{ij})$ is mapped to some element Δ. We now consider the bijection

$$B \otimes_A B \to \mathrm{Hom}_A(\mathrm{Hom}_A(B, A), B)$$
$$b \otimes c \mapsto (\varphi \mapsto \varphi(b)c),$$

and define θ to be the image of Δ. We remark that a priori θ is an A-module homomorphism between $\mathrm{Hom}_A(B, A)$ and B, which both have B-module structures. It is in fact a B-module homomorphism, and is moreover an isomorphism by Scheja and Storch [35, Satz 3.3]. Defining $\eta = \theta^{-1}(1)$, we have that η determines a bilinear form, which we denote by w

$$B \otimes_A B \to A$$
$$b \otimes c \overset{w}{\mapsto} \eta(bc).$$

Proof of Proposition 6   We note that, when $y_1 = \ldots = y_n = 0$, $\mathrm{Spec}\,(B)$ is the fiber of F over 0, consisting of a discrete set of points. This corresponds to a disjoint union of schemes. If b and c lie in different components, then their product is zero. This implies that the bilinear form w decomposes into an orthogonal direct sum of forms over each factor in $F^{-1}(0)$. These factors correspond to EKL forms at each point in the fiber $F^{-1}(0)$, and in particular over $0 \in F^{-1}(0)$, we recover the EKL form $w_0^{EKL}(F)$.   □

The following theorem will allow us to relate the EKL forms at various points in the fiber $F^{-1}(0)$.

Theorem 5 (Harder's Theorem) *[36, VII.3.13] A family of symmetric bilinear forms over $\mathbb{A}_k^1$ is constant (respectively, has constant specialization to k-points) for characteristic not equal to 2 (resp. any k). In particular when char(k) $\neq$ 2, for any finite k[t]-module M, we have that the family of bilinear forms $M \times_{k[t]} M \to k[t]$ is pulled back from some bilinear form $N \times_k N \to k$ via the unique morphism of schemes $\mathbb{A}_k^1 \to Spec(k)$.*

Step 3:   We choose y so that $F|_{F^{-1}(y)}$ is étale. One may use the generalization of Exercise 6(2) as stated in Eq. (2.5), combined with Proposition 5 to see that

$$\sum_{x \in F^{-1}(y)} w_x^{EKL}(F) = \sum_{x \in F^{-1}(y)} \deg_x^{\mathbb{A}^1} F.$$

By Harder's theorem, we have that $\sum_{x \in F^{-1}(y)} w_x^{EKL}(F) = \sum_{x \in F^{-1}(0)} w_x^{EKL}(F)$, and by the local formula for degree, we see that

$$\sum_{x \in F^{-1}(y)} \deg_x^{\mathbb{A}^1} F = \deg^{\mathbb{A}^1} F = \sum_{x \in F^{-1}(0)} \deg_x^{\mathbb{A}^1} F.$$

Thus $\sum_{x \in F^{-1}(0)} w_x^{EKL}(F) = \sum_{x \in F^{-1}(0)} \deg_x^{\mathbb{A}^1} F$. Since $F|_{F^{-1}(0) \setminus \{0\}}$ is étale, we may iteratively apply the equality in Eq. (2.5) to cancel terms, leaving us with the local degree and EKL form at the origin:

$$w_0^{EKL}(F) = \deg_0^{\mathbb{A}^1} F.$$

Therefore by finite determinacy we recover the desired equality $w_0^{EKL}(f) = \deg_0^{\mathbb{A}^1}(f)$. This concludes the Proof of Theorem 4.

2.3.3 $\mathbb{A}^1$-*Milnor Numbers*

The following section is based off of joint work by Jesse Kass and Kirsten Wickelgren [11, §8]. A variety over a perfect field is generically smooth, although it may admit a singular locus where the dimension of the tangent space exceeds the dimension of the variety, for example a self-intersecting point on a singular elliptic curve. Singularities are generally difficult to study, although certain classes are more tractable than others. There is a particular class of singularities, called *nodes*, which are in some sense the most generic. If k is a field of characteristic not equal to 2, then a node is given by an equation $x_1^2 + \ldots + x_n^2 = 0$ over a separable algebraic closure $\overline{k}$.

Consider a point p on a hypersurface $\{f(x_1, \ldots, x_n) = 0\} \subseteq \mathbb{A}_k^n$. Fix values $a_1, \ldots, a_n$, and consider the family

$$f(x_1, \ldots, x_n) + a_1 x_1 + \ldots + a_n x_n = t,$$

parametrized over the affine t-line. This hypersurface bifurcates into nodes over $\overline{k}$. Given any hypersurface $g(x_1, \ldots, x_n)$ with a node at a k-rational point p, we define the *type* of the node as the element in $\mathrm{GW}(k)$ corresponding to the rank one form represented by the Hessian matrix at p:

$$\mathrm{type}(p) := \left\langle \frac{\partial^2 g}{\partial x_i \partial x_j}(p) \right\rangle.$$

In particular, we see that:

$$\mathrm{type}(x_1^2 + a x_2^2 = 0) := \langle a \rangle$$

$$\mathrm{type}\left(\sum_{i=1}^n a_i x_i^2 = 0\right) := \left\langle 2^n \prod_{i=1}^n a_i \right\rangle.$$

In the case where we have a node at p with $k(p) = L$, then L is separable over k [37, Exposé XV, Théorème 1.2.6], and we define the type of the node as the trace of the type over its residue field. In the examples above, this gives:

$$\mathrm{type}\left(\sum_{i=1}^n a_i x_i^2 = 0\right) := \mathrm{Tr}_{L/k}\left\langle 2^n \prod_{i=1}^n a_i \right\rangle.$$

Thus the type encodes the field of definition of the node, as well as its tangent direction. In the case where $k = \mathbb{R}$, we can visualize the possible $\mathbb{R}$-rational nodes in degree two as:

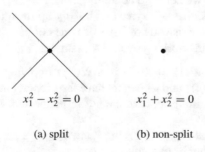

$$x_1^2 - x_2^2 = 0 \qquad\qquad x_1^2 + x_2^2 = 0$$

(a) split   (b) non-split

Here we may think of *split* as corresponding to the existence of rational tangent directions, while *non-split* refers to non-rational tangent directions. Over fields that aren't $\mathbb{R}$, it is possible to have many different split nodes.

In the case where $k = \mathbb{C}$, for any $(a_1, \ldots, a_n)$ sufficiently close to 0, it is a classical result that the number of nodes in this family is a constant integer, equal to $\deg_0^{top} \operatorname{grad} f =: \mu$, which is called the *Milnor number*. This admits a generalization as follows.

Theorem 6 *[12, Corollary 45] Assume that f has a single isolated singularity at the origin. Then for a generic $(a_1, \ldots, a_n)$, we have that the sum over nodes on the hypersurface $f + a_1 x_1 + \ldots + a_n x_n = t$ is*

$$\sum_{\substack{nodes\ p \\ in\ family}} type(p) = \deg_0^{\mathbb{A}^1} \operatorname{grad} f =: \mu_0^{\mathbb{A}^1} f.$$

We refer to this as the $\mathbb{A}^1$-Milnor number. We remark that the classical Milnor number can be recovered by taking the rank of the $\mathbb{A}^1$-Milnor number.

Example 6 Let $f(x, y) = x^3 - y^2$, over a field of characteristic not equal to 2 or 3. Let $p = (0, 0)$ be a point on the hypersurface $\{f = 0\}$. We can compute $\operatorname{grad} f = (3x^2, -2y)$, and then we have that

$$\deg \operatorname{grad} f = \deg(3x^2) \cdot \deg(-2y)$$

$$= \begin{pmatrix} 0 & 1/3 \\ 1/3 & 0 \end{pmatrix} \langle -2 \rangle$$

$$= \mathbb{H}.$$

This has rank two, so the classical Milnor number is $\mu = 2$. We can take our family to be $y^2 = x^3 + ax + t$. If $a = 0$, then we have a node at 0. In general, for $a \neq 0$, we have nodes at those t with the property that the discriminant of the curve $y^2 = x^3 + ax + t$ vanishes, that is at those t where $\Delta = -16(4a^3 + 27t^2) = 0$. This has at most two solutions in t, which we may denote by $\{x^2 + u_1 y^2 = 0\}$ and $\{x^2 + u_2 y^2 = 0\}$, and we see by Theorem 6 that $\mathbb{H} = \langle u_1 \rangle + \langle u_2 \rangle$. This implies, by taking determinants, that -1 agrees with $u_1 u_2$ up to squares. This provides us with obstructions to the existence of pairs of nodes of certain types, depending on the choice of field we are working over. For example:

- Over $\mathbb{F}_5$, we see that $\langle 1 \rangle = \langle -1 \rangle$ in $\operatorname{GW}(\mathbb{F}_5)$ implying that $u_1 u_2$ is always a square. In particular, u_1 and u_2 cannot have the property that exactly one of them is a non-square, meaning that we cannot bifurcate into a split and a non-split node.

- Over $\mathbb{F}_7$ we have that $\langle 1 \rangle \neq \langle -1 \rangle$, implying $u_1 u_2$ is a non-square, so we cannot bifurcate into two split or two non-split nodes.

Exercise 7 Compute $\mu^{\mathbb{A}^1}$ for the following ADE singularities over $\mathbb{Q}$:

Singularity	Equation
A_n	$x^2 + y^{n+1}$
D_n	$y(x^2 + y^{n-2})$ $(n \geq 4)$
E_6	$x^3 + y^4$
E_7	$x(x^2 + y^3)$
E_8	$x^3 + y^5.$

2.3.4   An Arithmetic Count of the Lines on a Smooth Cubic Surface

The following is based off of joint work of Jesse Kass and Kirsten Wickelgren [13]. Let $f \in k[x_0, x_1, x_2, x_3]$ be a homogeneous polynomial of degree three. Consider the following surface

$$V = \{f = 0\} \subseteq \operatorname{Proj} k[x_0, x_1, x_2, x_3] = \mathbb{P}^3_k,$$

and suppose that V is smooth.

Theorem 7 (Cayley-Salmon Theorem) *When $k = \mathbb{C}$, there are exactly 27 lines on V [38].*

Proof Consider the Grassmannian $\operatorname{Gr}_{\mathbb{C}}(2, 4)$, which parametrizes 2-dimensional complex subspaces $W \subseteq \mathbb{C}^{\oplus 4}$, or equivalently, lines in $\mathbb{P}^3_{\mathbb{C}}$. As the Grassmannian is a moduli space, it admits a *tautological bundle* S whose fiber over any point $W \in \operatorname{Gr}_{\mathbb{C}}(2, 4)$ is the vector space W itself. A chosen homogeneous polynomial f of degree three defines a section σ_f of $\operatorname{Sym}^3 S^*$, where

$$\sigma_f([W]) = f|_W.$$

Thus we see that the line $\ell \subseteq \mathbb{P}^3_{\mathbb{C}}$ corresponding to $[W]$ lies on the surface V if and only if $\sigma_f[W] = 0$. One may see that σ_f has isolated zeros [39, Corollary 6.17], and thus we may express the Euler class of the bundle as

$$e(\operatorname{Sym}^3 S^*) = c_4(\operatorname{Sym}^3 S^*) = \sum_\ell \deg_\ell^{\text{top}} \sigma_f, \tag{2.6}$$

where this last sum is over the zeros of σ_f. We determine $\deg_\ell^{\text{top}} \sigma_f$ by choosing local coordinates near ℓ on $\operatorname{Gr}_{\mathbb{C}}(2, 4)$ as well as a compatible trivialization for

$\mathrm{Sym}^3 \mathcal{S}^*$ over this coordinate patch. Then σ_f may be viewed as a function

$$\mathbb{A}_{\mathbb{C}}^4 \supseteq U \xrightarrow{\sigma_f} \mathbb{A}_{\mathbb{C}}^4$$

with an isolated zero at ℓ. We can then define $\deg_\ell^{\mathrm{top}} \sigma_f$ as the local degree of this function. It is a fact that the smoothness of V implies that σ_f vanishes to order 1 at ℓ. Thus the Euler class counts the number of lines on V. Finally, one may compute $c_4(\mathrm{Sym}^3 \mathcal{S}^*) = 27$ by applying the splitting principle and computing the cohomology of $\mathrm{Gr}_{\mathbb{C}}(2, 4)$.   □

In the real case, Schäfli [40] and Segre [41] showed that there can be 3, 7, 15, or 27 real lines on V. One of the main differences between the real and the complex case was the distinction that Segre drew between *hyperbolic* and *elliptic* lines.

Definition 5 We say that $I \in \mathrm{PGL}_2(\mathbb{R})$ is *hyperbolic* (resp. *elliptic*) if the set

$$\mathrm{Fix}(I) = \{x \in \mathbb{P}_{\mathbb{R}}^1 : Ix = x\}$$

consists of two real points (resp. a complex conjugate pair of points).

To a real line $\ell \subseteq V$ we may associate an involution $I \in \mathrm{Aut}(\ell) \cong \mathrm{PSL}_2(\mathbb{R})$, where I sends $p \in \ell$ to $q \in \ell$ if $T_p V \cap V = \ell \cup Q$, for some Q satisfying $\ell \cap Q = \{p, q\}$, (that is, for any point p on a line ℓ, there is exactly one other point q having the same tangent space). We can say that ℓ is hyperbolic (resp. elliptic) whenever I is.

Alternatively, we may describe these classes of lines topologically. We think of the frame bundle as a principal $\mathrm{SO}(3)$-bundle over $\mathbb{R}\mathbf{P}^3$. As $\mathrm{SO}(3)$ admits a double cover $\mathrm{Spin}(3)$, from any principal $\mathrm{SO}(3)$-bundle we may obtain a principal $\mathrm{Spin}(3)$-bundle. Traveling on our cubic surface along the line ℓ gives a distinguished choice of frame at every point on ℓ, that is, a loop in the frame bundle. This loop may or may not lift to the associated $\mathrm{Spin}(3)$-bundle. If the loop lifts, then ℓ is hyperbolic, and if it doesn't then ℓ is elliptic.

Theorem 8 *In the real case, we have the following relationship between hyperbolic and elliptic lines:*

$$\#\{real\ hyperbolic\ lines\ on\ V\} - \#\{real\ elliptic\ lines\ on\ V\} = 3.$$

We refer the reader to the following sources [41–45].

Proof sketch Via the map $\sigma_f : \mathrm{Gr}_{\mathbb{R}}(2, 4) \to \mathrm{Sym}^3 \mathcal{S}^*$, we have that

$$e(\mathrm{Sym}^3 \mathcal{S}^*) = \sum_{\substack{\ell \in \mathrm{Gr}_{\mathbb{R}}(2,4) \\ \sigma_f(\ell)=0}} \deg_\ell^{\mathrm{top}} \sigma_f.$$

One may also show that

$$\deg_\ell^{\text{top}} \sigma_f = \begin{cases} 1 & \text{if } \ell \text{ is hyperbolic} \\ -1 & \text{if } \ell \text{ is elliptic,} \end{cases}$$

and compute that $e(\text{Sym}^3 S^*) = 3$ using the Grassmannian of oriented planes. □

To define a notion of hyperbolic and elliptic which holds in more generality, we introduce the *type* of a line. As before, we let $V \subseteq \mathbb{P}_k^3$ be a smooth cubic surface, and consider a closed point $\ell \in \text{Gr}_k(2, 4)$, with residue field $L = k(\ell)$. We can then view ℓ as a closed immersion

$$\ell \cong \mathbb{P}_L^1 \hookrightarrow \mathbb{P}_k^3 \otimes_k L.$$

Given such a line $\ell \subseteq V$, we again have an associated involution:

$$I = \begin{pmatrix} a & b \\ c & d \end{pmatrix} \in \text{PGL}_2(L).$$

Since I is an involution, its fixed points satisfy $\frac{az+d}{cz+d} = z$, from which we can see they are defined over the field $L\left(\sqrt{D}\right)$, where D is the discriminant of the subscheme $\text{Fix}(I) \subseteq \mathbb{P}_L^1$.

Definition 6 The *type* of a line ℓ is the element of $\text{GW}(k(\ell))$ given by

$$\text{type}(\ell) := \langle D \rangle = \langle ad - bc \rangle = \langle -1 \rangle \deg^{\mathbb{A}^1}(I).$$

We say a line is *hyperbolic* if $\text{type}(\ell) = \langle 1 \rangle$, and *elliptic* otherwise.

Theorem 9 *[13, Theorem 2] The number of lines on a smooth cubic surface is computed via the following weighted count*

$$\sum_{\ell \subseteq V} Tr_{k(\ell)/k}(\text{type}(\ell)) = 15 \cdot \langle 1 \rangle + 12 \cdot \langle -1 \rangle .$$

Remark 5 We may apply the previous theorem to observe the following results:

1. If $k = \mathbb{C}$, then by taking the rank, we obtain the Cayley-Salmon Theorem (7), stating that the number of lines on a cubic surface is 27.
2. If $k = \mathbb{R}$, then $Tr_{\mathbb{C}/\mathbb{R}} \langle 1 \rangle = \langle 1 \rangle \oplus \langle -1 \rangle$. Taking the signature, we recover Theorem 8, stating that the number of hyperbolic lines minus the number of elliptic lines is 3.

As a particular application, if we are working over a finite field $k = \mathbb{F}_q$, then its square classes are $\mathbb{F}_q^\times \big/ (\mathbb{F}_q^\times)^2 \cong \{1, u\}$. Thus the type of a line ℓ over $\mathbb{F}_{q^a}$ is either $\langle 1 \rangle$ or $\langle u_a \rangle$, which by Definition 6 we call hyperbolic or elliptic, respectively.

Corollary 2 *[13, Theorem 1] For any natural number a, we have that the number of lines on V satisfies*

$$\#\{elliptic\ lines\ with\ field\ of\ definition\ \mathbb{F}_{q^{2a+1}}\}$$

$$+ \#\{hyperbolic\ lines\ with\ field\ of\ definition\ \mathbb{F}_{q^{2a}}\} \equiv 0 \pmod 2.$$

In particular when all the lines in question are defined over a common field k, we have that the number of elliptic lines is even.

In order to prove Theorem 9, one considers σ_f to be a section of the bundle $\mathrm{Sym}^3 S^* \to \mathrm{Gr}_k(2, 4)$, and computes a sum over its isolated zeros, weighted by their local index. Over the complex numbers, this is precisely Eq. (2.6), which recovers the Euler number of the bundle. In a more general context, however, we will want to obtain an element of $\mathrm{GW}(k)$. This requires us to use an enriched notion of an *Euler class*, described below.

Digression In this exposition, given a vector bundle $E \to X$ with section σ, we use the Euler class $e(E, \sigma)$ valued in $\mathrm{GW}(k)$ of [13, Section 4]. In the literature, there are a number of other Euler classes which coincide with this definition in various settings. One may define this Euler class via Chow-Witt groups [46] or oriented Chow groups [47] as in the work of M. Levine [48]. In his seminal book, Morel defines the Euler class of a bundle $E \to X$ as a cohomology class in twisted Milnor-Witt K-theory $H^n(X; \mathcal{K}_n^{\mathrm{MW}}(\det E^*))$ [16], and when $\det(E^*)$ is trivial, one may relate these Euler classes up to a unit multiple via the isomorphism

$$H^n(X; \mathcal{K}_n^{\mathrm{MW}}(\det E^*)) \cong \widetilde{\mathrm{CH}}(X, \det E^*).$$

For more details, see the work of Asok and Fasel [49]. Other versions of the Euler class in $\mathbb{A}^1$-homotopy theory occur in the work of Déglise, Jin and Khan [50] and the work of Levine and Raksit [51]. Many of these notions are equated in work of Bachmann and Wickelgren [52].

Definition 7 Let X be a smooth projective scheme of dimension r, and let $\mathcal{E} \to X$ be a rank r bundle. We say that $\mathcal{E}$ is *relatively oriented* if we are given an isomorphism

$$\mathrm{Hom}(\det TX, \det \mathcal{E}) \cong \mathcal{L}^{\otimes 2},$$

where $\mathcal{L}$ is a line bundle on X.

Suppose that σ is a section of a relatively oriented bundle $\mathcal{E}$ with isolated zeros, and define $Z = \{\sigma = 0\}$ to be its vanishing locus. For each $x \in Z$, we will define $\deg_x^{\mathbb{A}^1} \sigma$ as follows:

1. Choose *Nisnevich coordinates* [13, Definition 17] near $x \in Z$, that is, pick an open neighborhood $U \subseteq X$ around x, and an étale morphism $\varphi : U \to \mathbb{A}_k^r$ such that $k(\varphi(x)) \cong k(x)$.
2. Choose a *compatible oriented* trivialization $\mathcal{E}|_U$, that is, a local trivialization

$$\psi : \mathcal{E}|_U \to O_U^{\oplus r},$$

such that the associated section $\mathrm{Hom}(\det TX, \det \mathcal{E})(U)$ is a square of a section in $\mathcal{L}(U)$. Then we have that $\psi \circ \sigma \in O_U^{\oplus r}$ and there exists a $g \in (\mathfrak{m}_x^N)^{\oplus r}$, with N sufficiently large, so that

$$\psi \circ \sigma + g \in \varphi^* O_{\mathbb{A}_k^r}.$$

Define $f := \psi \circ \sigma + g$, and then we have that $f : \varphi(U) \to \mathbb{A}_k^r$ has an isolated zero at $\varphi(x)$. Since our trivialization was compatibly oriented, this definition is independent of the choice of g.
3. Finally, we define $\deg_x^{\mathbb{A}^1} \sigma := \deg_{\varphi(x)}^{\mathbb{A}^1} f \in \mathrm{GW}(k)$.

Definition 8 For a relatively oriented bundle $\mathcal{E} \to X$, and a section σ with isolated zeros, we define the *Euler class* to be

$$e(E, \sigma) := \sum_{x : \sigma(x) = 0} \deg_x^{\mathbb{A}^1} \sigma.$$

In order to conclude the proof of Theorem 9, we must identify $\deg_\ell^{\mathbb{A}^1} \sigma_f$ with type(ℓ). Then we are able to compute $e(\mathrm{Sym}^3 \mathcal{S}^*)$ using a well-behaved choice of cubic surface, for instance the Fermat cubic. For more details, see [13, §5].

Remark 6 Following our definition of an Euler class for a relatively oriented bundle, we include the following closely related remarks.

1. Interesting enumerative information is still available when relative orientability fails. For an example of this in the literature, we refer the reader to the paper of Larson and Vogt [53] which defines relatively oriented bundles relative to a divisor in order to compute an enriched count of bitangents to a smooth plane quartic [53].
2. Given a smooth projective scheme over a field, one may push forward the Euler class of its tangent bundle to obtain an Euler characteristic which is valued in $\mathrm{GW}(k)$. A particularly interesting consequence of this is an enriched version of the Riemann–Hurwitz formula, first established by M. Levine [48, Theorem 12.7] and expanded upon by work of Bethea, Kass, and Wickelgren [54].

Forthcoming work of Pauli investigates the related question of lines on quintic threefold [55]. We also refer the reader to work of M. Levine, which includes an examination of Witt-valued characteristic classes, including an Euler class of $\mathrm{Sym}^{2n-d} S^*$ on $\mathrm{Gr}_k(2, n+1)$ [56], and results of Bachmann and Wickelgren for symmetric bundles on arbitrary Grassmannians [52, Corollary 6.2]. Finally, for a further investigation of enriched intersection multiplicity, we refer the reader to recent work of McKean on enriching Bézout's Theorem [57].

2.3.5 An Arithmetic Count of the Lines Meeting 4 Lines in Space

The following is based off of work by Padmavathi Srinivasan and Kirsten Wickelgren [14].

In enumerative geometry, one encounters the following classical question: given four complex lines in general position in $\mathbb{C}P^3$, how many other complex lines meet all four? The answer is two lines, whose proof we sketch out below.

Four Lines in Three-Space, Classically Let L_1, L_2, L_3, L_4 be lines in $\mathbb{C}P^3$ so that no three of them intersect at one point (we refer to this condition as *general*). Given a point $p \in L_1$, there is a unique line L_p through p which intersects both L_2 and L_3. We then examine the surface swept out by all such lines $Q := \bigcup_{p \in L_1} L_p$, and we claim that this is a degree two hypersurface which contains L_1, L_2, and L_3. To see this, it suffices to verify that it is the vanishing locus of a degree two homogeneous polynomial. A homogeneous polynomial of degree two, considered as an element of $H^0(\mathbb{C}P^3, O(2))$, will vanish on the line L_i if and only if it lies in the kernel

$$H^0(\mathbb{C}P^3, O(2)) \to H^0(L_i, O(2)).$$

We verify that

$$\dim_k H^0(\mathbb{C}P^3, O(2)) = \binom{2+3}{2} = 10$$

$$\dim_k H^0(L_i, O(2)) = 3,$$

therefore for $i = 1, 2, 3$ each such map has kernel of dimension ≥ 7. This implies there is a polynomial f in the common kernel of all three maps. We claim that $L_p \subseteq V(f)$ for each $p \in L_1$, and indeed since three points of L_p lie in $V(f)$, we see that $V(f)$ contains the entire line. Therefore we have containment $V(f) \supseteq Q$, and it is easy to see we must have equality. Finally by applying Bézout's Theorem, we see that $Q \cap L_4$ consists of two points, counted with multiplicity.

One might ask how to answer this question over an arbitrary field k. We recall that the Grassmannian $\mathrm{Gr}_k(2, 4)$ parametrizes lines in $\mathbb{P}_k^3$ (that is, two-dimensional subspaces of $k^{\oplus 4}$), which is an appealing moduli space for this problem. We first select a basis $\{e_1, e_2, e_3, e_4\}$ of $k^{\oplus 4}$ satisfying

$$L_1 = ke_3 \oplus ke_4,$$

and we define a new line L such that

$$L = k\widetilde{e}_3 \oplus k\widetilde{e}_4,$$

where $\widetilde{e}_3$ and $\widetilde{e}_4$ are some linearly independent vectors whose definition we defer until further below. Letting φ_i denote the dual basis element to e_i, one may compute that $L \cap L_1$ is nonempty if and only if

$$(\varphi_1 \wedge \varphi_2)(\widetilde{e}_3 \wedge \widetilde{e}_4) = 0.$$

Consider the line bundle $\det S^* = S^* \wedge S^* \to \mathrm{Gr}_k(2, 4)$, whose fiber over a point $W \in \mathrm{Gr}_k(2, 4)$ is $W^* \wedge W^*$. We then have that $\varphi_1 \wedge \varphi_2 \in H^0(\mathrm{Gr}_k(2, 4), S^* \wedge S^*)$ and

$$(\varphi_1 \wedge \varphi_2)([W]) = \varphi_1|_W \wedge \varphi_2|_W .$$

It is then clear that we obtain a bijection between lines intersecting L_1 and zeros of $\varphi_1 \wedge \varphi_2$:

$$\{L \ : \ L \cap L_1 \neq \varnothing\} = \{[W] \ : \ (\varphi_1 \wedge \varphi_2)([W]) = 0\}.$$

We may repeat this process for each line to form a section σ of $\oplus_{i=1}^4 S^* \wedge S^*$. Then the zeros of σ will correspond exactly to lines which meet all four of our chosen lines:

$$\{L \ : \ L \cap L_i \neq \varnothing, \ i = 1, 2, 3, 4\} = \{[W] \in \mathrm{Gr}_k(2, 4) \ : \ \sigma([W]) = 0\}.$$

In particular, if σ is a section of a relatively oriented bundle, then we may calculate an enriched count of lines meeting four lines in space, given by the Euler class

$$e\left(\oplus_{i=1}^4 S^* \wedge S^*, \sigma\right) = \sum_{L \ : \ L \cap L_i \neq 0} \mathrm{ind}_L \sigma. \tag{2.7}$$

Denote by $\mathcal{E} = \oplus_{i=1}^4 S^* \wedge S^*$ our rank four vector bundle over $X := \mathrm{Gr}_k(2, 4)$. Since X is a smooth projective scheme of dimension four, we have that $(\det TX)^* \cong \omega_X \cong O(-2)^{\otimes 2}$, and $\det \mathcal{E} \cong \left(\otimes_{i=1}^2 S^* \wedge S^*\right)^{\otimes 2}$. Therefore $\mathrm{Hom}(\det TX, \det \mathcal{E}) \cong \omega_X \otimes \det \mathcal{E} \cong \mathcal{L}^{\otimes 2}$, so $\mathcal{E}$ is relatively oriented over X, and Eq. (2.7) is a valid expression. In order to compute a local index of the section σ near a zero L, we must first parametrize Nisnevich local coordinates near L. Here we define a parametrized basis of $k^{\oplus 4}$ by

$$\widetilde{e}_1 = e_1$$
$$\widetilde{e}_2 = e_2$$
$$\widetilde{e}_3 = xe_1 + ye_2 + e_3$$
$$\widetilde{e}_4 = x'e_1 + y'e_2 + e_4.$$

We then obtain a morphism from affine space to an open cell around L:

$$\mathbb{A}_k^4 = \operatorname{Spec} k[x, y, x', y'] \to U \subseteq \operatorname{Gr}_k(2, 4)$$
$$(x, y, x', y') \mapsto \operatorname{span}\{\widetilde{e}_3, \widetilde{e}_4\}.$$

Over this cell, we obtain an oriented trivialization of the bundle $\det S^*$, given by $\widetilde{\varphi}_3 \wedge \widetilde{\varphi}_4$, where $\widetilde{\varphi}_i$ denotes the dual basis element to $\widetilde{e}_i$. Under these local coordinates, we may compute the local index $\operatorname{ind}_L \sigma$ as the local $\mathbb{A}^1$-degree at the origin of the induced map $\mathbb{A}_k^4 \to \mathbb{A}_k^4$. Suppose that

$$L_1 = \{\varphi_1 = \varphi_2 = 0\} = ke_3 \oplus ke_4.$$

Then we have that $\sigma([W]) = \left(\varphi_1 \wedge \varphi_2|_{[W]}, \ldots\right)$. We see then that

$$(\varphi_1 \wedge \varphi_2)|_{k\widetilde{e}_3 \oplus k\widetilde{e}_4} = (x\widetilde{\varphi}_3 + y\widetilde{\varphi}_4) \wedge (x'\widetilde{\varphi}_3 + y'\widetilde{\varphi}_4)$$
$$= (xy' - x'y)\widetilde{\varphi}_3 \wedge \widetilde{\varphi}_4.$$

Thus we may exhibit σ as a function

$$f = (f_1, f_2, f_3, f_4) : \mathbb{A}_k^4 \to \mathbb{A}_k^4,$$

where $f_1(x, y, x', y') = xy' - x'y$. Then in the basis (x, y, x', y') we have that the Jacobian of σ has its first column as:

$$\operatorname{Jac}(f) = \det \begin{pmatrix} y' & \cdots \\ -x' & \cdots \\ -y & \cdots \\ x & \cdots \end{pmatrix}.$$

Question Is there a geometric interpretation of $\operatorname{ind}_L \sigma = \deg_L^{\mathbb{A}^1} f$?

The intersections $L \cap L_i$ for $i = 1, \ldots, 4$ determine four points on $L \cong \mathbb{P}_{k(L)}^1$. Let λ_L denote the cross-ratio of these points in $k(L)^*$. Denote by P_i the plane spanned by L and L_i. We note that planes P in $\mathbb{P}_k^3$ correspond to subspaces $V \subseteq k(P)^{\oplus 4}$

where $\dim(V) = 3$. If P contains the line $L = [W]$ then it corresponds to $W \subseteq V \subseteq$ $k(P)^{\oplus 4}$, which in turn corresponds to $k(P)$-points of $\text{Proj}\left(k(L)^{\oplus 4}\big/ W\right) \cong \mathbb{P}^1_{k(L)}$. Thus we might think of the planes P_i for $i = 1, \ldots, 4$ as 4 points on $\mathbb{P}^1_{k(L)}$. Let μ_L denote the cross-ratio of these points.

Theorem 10 *[14, Theorem 1] Let L_1, L_2, L_3, L_4 be four general lines defined over k in $\mathbb{P}^3_k$. Then*

$$\sum_{\{L \, : \, L \cap L_i \neq \varnothing \; \forall i\}} Tr_{k(L)/k} \langle \lambda_L - \mu_L \rangle = \langle 1 \rangle + \langle -1 \rangle.$$

As a generalization, let $\pi_1, \ldots, \pi_{2n-2}$ be codimension 2 planes in $\mathbb{P}^n_k$ for n odd. Then

$$\sum_{\{L \, : \, L \cap \pi_i \neq \varnothing \; \forall i\}} Tr_{k(L)/k} \det \begin{pmatrix} \cdots & c_i b_1^i & \cdots \\ \cdots & c_i b_2^i & \cdots \end{pmatrix} = \frac{1}{2n} \binom{2n-2}{n-1} \mathbb{H},$$

where c_i are normalized coordinates for the line $\pi_i \cap L$ (defined in [14, Definition 10]), and $[b_1^i, b_2^i] = L \cap \pi_i \cong \mathbb{P}^1_{k(L)}$. This weighted count is expanded in forthcoming work of the author, which provides a generalized enriched count of m-planes meeting mp codimension m planes in $(m + p)$-space [58].

Corollary 3 *[14, Corollary 3] Over $\mathbb{F}_q$, we cannot have a line L over $\mathbb{F}_{q^2}$ with*

$$\lambda_L - \mu_L = \begin{cases} non\text{-}square & q \equiv 3 \pmod 4 \\ square & q \equiv 1 \pmod 4. \end{cases}$$

For related results in the literature, we refer the reader to the papers of Levine and Bachmann–Wickelgren mentioned in the previous section [52, 56], as well as Wendt's work developing a Schubert calculus valued in Chow-Witt groups [59]. Finally, Pauli uses Macaulay2 to compute enriched counts over a finite field of prime order and the rationals for various problems presented in these conference proceedings, including lines on a cubic surface, lines meeting four general lines in space, the EKL class, and various $\mathbb{A}^1$-Milnor numbers [60].

Notation Guide

$\langle a \rangle$	The element of the Grothendieck–Witt ring corresponding to $a \in k^\times/(k^\times)^2$
$[X, Y]_{\mathbb{A}^1}$	Genuine $\mathbb{A}^1$-homotopy classes of morphisms $X \to Y$
$\text{Béz}(f/g)$	Bézout bilinear form of a rational function

$\mathrm{Bl}_Z X$	Blowup of a subscheme Z in X
$\deg^{\mathbb{A}^1}$	Global $\mathbb{A}^1$-degree
$\deg^{\mathrm{top}}$	The topological (Brouwer) degree of a map between real or complex manifolds
$e(\mathcal{F})$	The Euler number (Euler class) of a vector bundle
$\mathbb{G}_m$	The multiplicative group scheme
$\mathrm{Gr}_k(n, m)$	The Grassmannian of affine n-planes in m-space over a field k
$\mathrm{GW}(k)$	The Grothendieck–Witt ring over a field k
$\mathbb{H}$	The hyperbolic element $\langle 1 \rangle + \langle -1 \rangle$ in $\mathrm{GW}(k)$
$\mu^{\mathbb{A}^1}$	$\mathbb{A}^1$-Milnor number
$[X, Y]_N$	Naive $\mathbb{A}^1$-homotopy classes of morphisms $X \to Y$
$N_Z X$	The normal bundle of a subscheme Z in X
$\mathrm{Sh}_\tau(\mathcal{C})$	The category of sheaves in a Grothendieck topology τ on a category $\mathcal{C}$
$\mathrm{sPre}(\mathcal{C})$	The category of simplicial presheaves on $\mathcal{C}$
$\mathrm{Th}(V)$	Thom space of a vector bundle V
$\mathrm{Tr}_{L/K}$	Trace for a field extension L/K
w^{EKL}	The Eisenbud-Levine/Khimshiashvili bilinear form

Acknowledgments This expository paper is based around lectures by and countless conversations with Kirsten Wickelgren, who introduced me to this area of mathematics and has provided endless guidance and support along the journey. I am grateful to Mona Merling, who has shaped much of my mathematical understanding, and to Stephen McKean and Sabrina Pauli for many enlightening mathematical discussions related to $\mathbb{A}^1$-enumerative geometry. I am also grateful to Frank Neumann and Ambrus Pál for their work organizing these conference proceedings. Finally, I would like to thank the anonymous referee for their thoughtful feedback, which greatly improved this paper. The author is supported by an NSF Graduate Research Fellowship, under grant number DGE-1845298.

References

1. M. Karoubi, O. Villamayor, K-théorie algébrique et K-théorie topologique, I. Math. Scand. **28**, 265–307 (1971/1972). ISSN: 0025-5521. https://doi.org/10.7146/math.scand.a-11024.
2. C.A. Weibel, Homotopy algebraic K4-theory, in *Algebraic K-theory and Algebraic Number Theory (Honolulu, HI, 1987)*, vol. 83. Contemporary Mathematics (American Mathematical Society, Providence, RI, 1989), pp. 461–488. https://doi.org/10.1090/conm/083/991991
3. J.F. Jardine, Algebraic homotopy theory, Canadian J. Math. **33**(2), 302–319 (1981). ISSN: 0008-414X. https://doi.org/10.4153/CJM-1981-025-9
4. J.F. Jardine, Algebraic homotopy theory and some homotopy groups of algebraic groups, C. R. Math. Rep. Acad. Sci. Canada **3**(4), 191–196 (1981). ISSN: 0706-1994
5. A. Suslin, V. Voevodsky, Singular homology of abstract algebraic varieties. Invent. Math. **123**(1), 61–94 (1996). issn: 0020-9910. https://doi.org/10.1007/BF01232367
6. V. Voevodsky, $\mathbf{A}^1$-homotopy theory, in *Proceedings of the International Congress of Mathematicians (Berlin, 1998)*, vol. I. (1998), pp. 579–604

7. D.G. Quillen. *Homotopical Algebra*. Lecture Notes in Mathematics, vol. 43 (Springer, New York, 1967), pp. iv+156 (not consecutively paged)
8. F. Morel, Théorie homotopique des schémas. Astérisque **256**, vi+119 (1999). ISSN: 0303-1179
9. F. Morel, V. Voevodsky, $\mathbb{A}^1$-homotopy theory of schemes. Inst. Hautes Études Sci. Publ. Math. **90**, 45–143 (1999/2001). ISSN: 0073-8301. http://www.numdam.org/item?id=PMIHES_ 1999__90__45_0
10. V. Voevodsky, On motivic cohomology with **Z**/l-coefficients. Ann. of Math. (2) **174**(1), 401– 438 (2011). ISSN: 0003-486X. https://doi.org/10.4007/annals.2011.174.1.11
11. J.L. Kass, K. Wickelgren, The class of Eisenbud-Khimshiashvili-Levine is the local $\mathbb{A}^1$- Brouwer degree. Duke Math. J **168**(3), 429–469 (2019). ISSN: 0012-7094. https://doi.org/10. 1215/00127094-2018-0046
12. J. Kass, K. Wickelgren. *A1-Milnor Number* (2016). https://services.math.dukc.cdu/~kgw/ papers/owr-Kass-Wickelgren.pdf.
13. J.L. Kass, K. Wickelgren, An Arithmetic Count of the Lines on a Smooth Cubic Surface (2017). arXiv preprint arXiv:1708.01175
14. P. Srinivasan, K. Wickelgren, An arithmetic count of the lines meeting four lines in $\mathbb{P}^3$ (2018). arXiv preprint arXiv:1810.03503
15. F. Morel, $\mathbb{A}^1$-algebraic topology, in *International Congress of Mathematicians*, vol. II. (European Mathematical Society, Zürich, 2006), pp. 1035–1059
16. F. Morel, $\mathbb{A}^1$-*algebraic Topology Over a Field*, vol. 2052. Lecture Notes in Mathematics (Springer, Heidelberg, 2012), pp. x+259. ISBN: 978-3-642-29513-3. http://dx.doi.org/10.1007/ 978-3-642-29514-0.
17. T.Y. Lam. *Introduction to Quadratic Forms Over Fields*, vol. 67. Graduate Studies in Mathematics (American Mathematical Society, Providence, RI, 2005), pp. xxii+550. ISBN: 0-8218-1095-2
18. J.L. Kass, K. Wickelgren, A classical proof that the algebraic homotopy class of a rational function is the residue pairing, Linear Algebra Appl. **595**, 157–181 (2020). ISSN: 0024-3795. https://doi.org/10.1016/j.laa.2019.12.041
19. C. Cazanave, Algebraic homotopy classes of rational functions. Ann. Sci. Éc. Norm. Supér. (4) **45**(4), 511–534 (2012/2013). ISSN: 0012-9593. https://doi.org/10.24033/asens.2172
20. A. Asok, *Algebraic Geometry from an A1-homotopic Viewpoint* (2019). https://dornsife.usc. edu/assets/sites/1176/docs/PDF/MATH599.S16.pdf
21. T. Lawson, *An Introduction to Bousfield Localization* (2020). arXiv: 2002.03888 [math.AT]
22. D. Dugger, S. Hollander, D.C. Isaksen, Hypercovers and simplicial presheaves, Math. Proc. Cambridge Philos. Soc. **136**(1), 9–51 (2004). ISSN: 0305-0041. https://doi.org/10.1017/ S0305004103007175.
23. B. Antieau, E. Elmanto, A primer for unstable motivic homotopy theory, in *Surveys on Recent Developments in Algebraic Geometry*, vol. 95. Proceedings Symposium Pure Mathematical (American Mathematical Society, Providence, RI, 2017), pp. 305–370
24. D. Dugger, Universal homotopy thcories. Adv. Math. **164**(1), 144–176 (2001). ISSN: 0001- 8708. https://doi.org/10.1006/aima.2001.2014
25. T. Bachmann, M. Hoyois, Norms in motivic homotopy theory (2017). arXiv preprint arXiv:1711.03061
26. M. Robalo, K-theory and the bridge from motives to noncommutative motives, Adv. Math. **269**, 399–550 (2015). ISSN: 0001-8708. https://doi.org/10.1016/j.aim.2014.10.011.
27. S. Bosch, W. Lütkebohmert, M. Raynaud, *Néron models*, vol. 21. Ergebnisse der Mathematik und ihrer Grenzgebiete (3) [Results in Mathematics and Related Areas (3)] (Springer, Berlin, 1990), pp. x+325. ISBN: 3-540-50587-3. https://doi.org/978-3-642-51438-8
28. W. Fulton, *Intersection Theory*, 2nd edn., vol. 2. Ergebnisse der Mathematik und ihrer Grenzgebiete. 3. Folge. A Series of Modern Surveys in Mathematics [Results in Mathematics and Related Areas. 3rd Series. A Series of Modern Surveys in Mathematics]. (Springer, Berlin, 1998), pp. xiv+470. ISBN: 3-540-62046-X; 0-387-98549-2. https://doi.org/10.1007/978-1- 4612-1700-8

29. M. Hoyois, The six operations in equivariant motivic homotopy theory Adv. Math. **305**, 197–279 (2017). ISSN: 0001-8708. https://doi.org/10.1016/j.aim.2016.09.031.
30. M. Hoyois, A quadratic refinement of the Grothendieck-Lefschetz-Verdier trace formula. Algebr. Geom. Topol. **14**(6), 3603–3658 (2014). ISSN: 1472-2747. http://dx.doi.org.prx. library.gatech.edu/10.2140/agt.2014.14.3603.
31. T. Brazelton et al., The trace of the local $\mathbf{A}^1$-degree. Homology Homotopy Appl. **23**(1), 243–255 (2020). arXiv: 1912.04788 [math.AT]
32. D. Eisenbud, H.I. Levine, An algebraic formula for the degree of a C^∞ map germ, Ann. Math. (2) **106**(1), 19–44 (1977). With an appendix by Bernard Teissier, *Sur une inégalité à la Minkowski pour les multiplicités*, pp. 19–44. https://doi.org/10.2307/1971156
33. G.N. Himšiašvili, The local degree of a smooth mapping, Sakharth. SSR Mecn. Akad. Moambe **85**(2), 309–312 (1977)
34. V.P. Palamodov, The multiplicity of a holomorphic transformation, Funkcional. Anal. i Priložen. **1**(3), 54–65 (1967). ISSN: 0374-1990
35. G. Scheja, U. Storch, Über Spurfunktionen bei vollständigen Durchschnitten, in *Journal für die reine und angewandte Mathematik* 0278_0279 (1975), pp. 174–190. http://eudml.org/doc/151652
36. T.Y. Lam. *Serre's problem on projective modules*. Springer Monographs in Mathematics (Springer, Berlin, 2006), pp. xxii+401. ISBN: 978-3-540-23317-6; 3-540-23317-2. https://doi.org/10.1007/978-3-540-34575-6
37. P. Deligne, N. Katz, *Groupes de Monodromie en géométrie algébrique. II*. Lecture Notes in Mathematics, vol. 340. Séminaire de Géométrie Algébrique du Bois-Marie 1967-1969 (SGA 7 II) (Springer, New York, 1973), pp. x+438
38. A. Cayley, On the triple tangent planes of surfaces of the third order, in *sThe Collected Mathematical Papers*, vol. 1. Cambridge Library Collection—Mathematics (Cambridge University, Cambridge, 2009), pp. 445–456. https://doi.org/10.1017/CBO9780511703676.077
39. D. Eisenbud, J. Harris, *3264 and all that—a Second Course in Algebraic Geometry* (Cambridge University, Cambridge, 2016), pp. xiv+616. ISBN: 978-1-107-60272-4; 978-1-107-01708-5. https://doi.org/10.1017/CBO9781139062046.
40. L. Schäfli, Ueber eine symbolische Formel, die sich auf die Zusammensetzung der binären quadratischen Formen bezieht, J. Reine Angew. Math. **57**, 170–174 (1860). ISSN: 0075-4102. https://doi.org/10.1515/crll.1860.57.170.
41. B. Segre. *The Non-singular Cubic Surfaces* (Oxford University, Oxford, 1942), pp. xi+180
42. R. Benedetti, R. Silhol, Spin and Pin$^-$ structures, immersed and embedded surfaces and a result of Segre on real cubic surfaces. Topology **34**(3), 651–678 (1995). ISSN: 0040-9383. https://doi.org/10.1016/0040-9383(94)00046-N.
43. A. Horev, J.P. Solomon, The open Gromov-Witten-Welschinger theory of blowups of the projective plane (2012). arXiv preprint arXiv:1210.4034
44. C. Okonek, A. Teleman, Intrinsic signs and lower bounds in real algebraic geometry. J. Reine Angew. Math. **688**, 219–241 (2014). ISSN: 0075-4102. https://doi.org/10.1515/crelle-2012-0055
45. S. Finashin, V. Kharlamov, Abundance of 3-planes on real projective hypersurfaces, Arnold Math. J. **1**(2), 171–199 (2015). ISSN: 2199-6792. https://doi.org/10.1007/s40598-015-0015-5
46. J. Barge, F. Morel, Groupe de Chow des cycles orientés et classe d'Euler des fibrés vectoriels. C. R. Acad. Sci. Paris Sér. I Math. **330**(4), 287–290 (2000). ISSN: 0764-4442. https://doi.org/10.1016/S0764-4442(00)00158-0
47. J. Fasel, Groupes de Chow-Witt, Mém. Soc. Math. Fr. (N.S.) **113**, viii+197 (2008). ISSN: 0249-633X. https://doi.org/10.24033/msmf.425
48. M. Levine, Toward an enumerative geometry with quadratic forms (2017). arXiv: 1703.03049 [math.AG]
49. A. Asok, J. Fasel, Comparing Euler classes. Q. J. Math **67**(4), 603–635 (2016). ISSN: 0033-5606
50. F. Déglise, F. Jin, A. Khan, *Fundamental Classes in Motivic Homotopy Theory* (2018), preprint. https://arxiv.org/abs/1805.05920

51. M. Levine, A. Raksit, Motivic Gauß-Bonnet Formulas (2018), preprint. https://arxiv.org/abs/1808.08385
52. T. Bachmann, K. Wickelgren, $\mathbb{A}^1$-*Euler Classes: Six Functors Formalisms, Dualities, Integrality and Linear Subspaces of Complete Intersections* (2020). arXiv: 2002.01848 [math.KT]
53. H. Larson, I. Vogt, *An Enriched Count of the Bitangents to a Smooth Plane Quartic Curve* (2019). arXiv: 1909.05945 [math.AG]
54. C. Bethea, J.L. Kass, K. Wickelgren, Examples of wild ramification in an enriched Riemann-Hurwitz formula, in *Motivic Homotopy Theory and Refined Enumerative Geometry*, vol. 745. Contemporary Mathematics (American Mathematical Society, Providence, RI, 2020), pp. 69–82. https://doi.org/10.1090/conm/745/15022 .
55. S. Pauli, *Quadratic Types and the Dynamic Euler Number of Lines on a Quintic Threefold* (2020). arXiv: 2006.12089 [math.AG]
56. M. Levine, Motivic Euler characteristics and Witt-valued characteristic classes. Nagoya Math. J. **236**, 251–310 (2019). ISSN: 0027-7630. https://doi.org/10.1017/nmj.2019.6.
57. S. McKean, *An Arithmetic Enrichment of Bézout's Theorem* (2020). arXiv: 2003.07413 [math.AG]
58. T. Brazelton, *An Enriched Degree of the Wronski*. In preparation (2020)
59. M. Wendt, Oriented Schubert calculus in Chow-Witt rings of Grassmannians (2018). arXiv preprint arXiv:1808.07296
60. S. Pauli, *Computing A1-Euler Numbers with Macaulay2* (2020). arXiv: 2003.01775 [math.AG]

Chapter 3
Cohomological Methods in Intersection Theory

Denis-Charles Cisinski

Abstract The goal of these notes is to see how motives may be used to enhance cohomological methods, giving natural ways to prove independence of ℓ results and constructions of characteristic classes (as 0-cycles). This leads to the Grothendieck-Lefschetz formula, of which we give a new motivic proof. There are also a few additions to what have been told in the lectures:

- A proof of Grothendieck-Verdier duality of étale motives on schemes of finite type over a regular quasi-excellent scheme (which slightly improves the level of generality in the existing literature).
- A proof that $\mathbf{Q}$-linear motivic sheaves are virtually integral (Theorem 3.3.2.12).
- A proof of the motivic generic base change formula.

3.1 Introduction

Let p be a prime number and $q = p^r$ a power of p. Let X_0 be a smooth and projective algebraic variety over $\mathbf{F}_q$. It comes equipped with the geometric Frobenius map $\phi_r : X \to X$, where $X = X_0 \times_{\mathbf{F}_q} \bar{\mathbf{F}}_p$, so that the locus of fixed points of F corresponds to the set of rational points of X_0 (various Frobenius morphisms and their actions are discussed in detail in Remark 3.4.2.24 below). We may take the graph of Frobenius $\Gamma_{\phi_r} \subset X \times X$, intersect with the diagonal, then interpret the intersection number cohomologically with the formula of Lefschetz through ℓ-adic cohomology, with $\ell \neq p$.

For each $Z \subseteq X$ we can attach a cycle $[Z] \in H^*(X, \mathbf{Q}_\ell)$ and do intersection theory (interpreting geometrically the algebraic operations on cycle classes). For instance, if $Z' \subseteq X$ is another cycle which is transversal to Z, we have $[Z] \cdot [Z'] = [Z \cap Z']$. Together with Poincaré duality, this implies that the number of rational

D.-C. Cisinski (✉)
Fakultät für Mathematik, Universität Regensburg, Regensburg, Germany
e-mail: denis-charles.cisinski@ur.de

© The Author(s), under exclusive license to Springer Nature Switzerland AG 2021   49
F. Neumann, A. Pál (eds.), *Homotopy Theory and Arithmetic Geometry – Motivic and Diophantine Aspects*, Lecture Notes in Mathematics 2292,
https://doi.org/10.1007/978-3-030-78977-0_3

points of X_0 may be computed cohomologically:

$$\#X(\mathbf{F}_q) = \sum_i (-1)^i \operatorname{Tr}\big(\phi_r^* : H^i(X, \mathbf{Q}_\ell) \to H^i(X, \mathbf{Q}_\ell)\big).$$

The construction of ℓ-adic cohomology by Grothendieck was aimed precisely at proving this kind of formulas, with the goal of proving Weil's conjectures on the ζ-functions of smooth and projective varieties over finite fields, which was finally achieved by Deligne [16, 18].

Here are two natural problems we would like to discuss:

- Extend this to non-smooth or non-proper schemes and to cohomology with possibly non-constant coefficients: this is what the Grothendieck-Lefschetz formula is about.
- Address the problem of independance on ℓ (when we compute traces of endomorphisms with a less obvious geometric meaning): this is what motives are made for.

In this series of lectures, I will explain what is a motive and explain how to prove a motivic Grothendieck-Lefschetz formula. To be more precise, we shall work with *h-motives* over a scheme X, which are one of the many descriptions of étale motives. These are the objects of the triangulated category $DM_h(X)$ constructed and studied in details in [13], which is a natural modification (the non-effective version) of an earlier construction of Voevodsky [47], following the lead of Morel and Voevodsky into the realm of $\mathbf{P}^1$-stable $\mathbf{A}^1$-homotopy theory of schemes. Although we will not mention them in these notes, we should mention that there are other equivalent constructions of étale motives which are discussed in [13] and [5] (not to speak of the many models with $\mathbf{Q}$-coefficients discussed in [14]), and more importantly, that there are also other flavours of motives [12, 28, 49], which are closer to geometry (and further from topology), for which one can still prove Lefschetz-Verdier formulas; see [27]. As we will see later, étale motives with torsion coefficients may be identified with classical étale sheaves. In particular, when restricted to the case of torsion coefficients, all the results discussed in these notes on trace formulas go back to Grothendieck [22]. The case of rational coefficients has also been studied previously to some extend by Olsson [34, 35]. We will see here how these fit together, as statements about étale motives with arbitrary (e.g. integral) coefficients. Finally, we will recall the Lefschetz-Verdier trace formula and explain how to deduce from it the motivic Grothendieck-Lefschetz formula, using Bondarko's theory of weights and Olsson's computations of local terms of the motivic Lefschetz-Verdier trace.

3.2 Étale Motives

3.2.1 The h-topology

Definition 3.2.1.1 A morphism of schemes $f : X \to Y$ is a *universal topological isomorphism* (*epimorphism* resp.) if for any map of schemes $Y' \to Y$, the pullback $X' = Y' \times_Y X \to Y'$ is a homeomorphism (a topological epimorphism resp., which means that it is surjective and exhibits Y' as a topological quotient).

Example 3.2.1.2 Surjective proper maps as well as faithfully flat maps all are universal epimorphisms.

Proposition 3.2.1.3 *A morphism of schemes $f : X \to Y$ is a universal homeomorphism if and only if it is surjective radicial and integral. Namely, f is integral and, for any algebraically closed field K, induces a bijection $X(K) \cong Y(K)$.*

Example 3.2.1.4 The map $X_{red} \to X$ is a universal homeomorphism.

Example 3.2.1.5 Let K'/K be a purely inseparable extension of fields. If X is a normal scheme with field of functions K, and if X' is the normalization of X in K', then the induced map $X' \to X$ is a universal homeomorphism.

Following Voevodsky [47], we can define the *h-topology* as the Grothendieck topology on noetherian schemes associated to the pre-topology whose coverings are finite families $\{X_i \to X\}_{i \in I}$ such that the induced map $\coprod_i X_i \to X$ is a universal epimorphism.[1] Beware that the h-topology is not subcanonical: any universal homeomorphism becomes invertible locally for the h-topology.

Using Raynaud-Gruson's flatification theorem, one shows the following; see [41].

Theorem 3.2.1.6 (Voevodsky, Rydh): *Let $X_i \to X$ be an h-covering. Then there exists an open Zariski cover $X = \cup_j X_j$ and for each j a blow-up $U'_j = Bl_{Z_j} U_j$ for some closed subset $Z_j \subseteq U_j$, a finite faithfully flat $U''_j \to U'_j$ and a Zariski covering $\{V_{j,\alpha}\}_\alpha$ of U''_j such that we have a dotted arrow making the following diagram commutative.*

$$
\begin{array}{ccc}
\coprod_{j,\alpha} V_{j,\alpha} & \cdots\cdots\cdots\cdots\cdots\cdots\cdots\cdots\cdots\cdots\longrightarrow & \coprod_i X_i \\
\downarrow & & \downarrow \\
\coprod_j U''_j \longrightarrow \coprod_j U'_j \longrightarrow \coprod_j U_j & \longrightarrow & X
\end{array}
$$

This means that the property of descent with respect to the h-topology is exactly the property of descent for the Zariski topology, together with proper descent.

---

[1] As shown by D. Rydh [41], this topology can be extended to all schemes, at the price of adding compatiblities with the constructible topology.

Remark 3.2.1.7 Although Grothendieck topologies where not invented yet, a significant amount of the results of SGA 1 [23] are about h-descent of étale sheaves (and this is one of the reasons why the very notion of descent was introduced in SGA 1). This goes on in SGA 4 [3] where the fact that proper surjective maps and étale surjective maps are morphism of universal cohomological descent is discussed at length. However, it is only in Voevodsky's thesis [47] that the h-topology is defined and studied properly, with the clear goal to use it in the definition of a triangulated category of étale motives.

3.2.2   Construction of Motives, After Voevodsky

3.2.2.1 Let Λ be a commutative ring. Let $Sh_h(X, \Lambda)$ denote the category of sheaves of Λ-modules on the category of separated schemes of finite type over X with respect to the h-topology. We have Yoneda functor

$$Y \mapsto \Lambda(Y),$$

where $\Lambda(Y)$ is the h-sheaf associated to the presheaf $\Lambda[\mathrm{Hom}_X(-, Y)]$ (the free Λ module generated by $\mathrm{Hom}_X(-, Y)$).

Let us consider the derived category $D(Sh_h(X, \Lambda))$, i.e. the localization of complexes of sheaves by the quasi-isomorphisms. Here we will speak the language of ∞-categories right away.[2] In particular, the word 'localization' has to be interpreted higher categorically (if we take as models simplicial categories, this is also known as the Dwyer-Kan localization). That means that $D(Sh_h(X, \Lambda))$ is in fact a stable ∞-category with small limits and colimits (as is any localization of a stable model category). Moreover, the constant sheaf functor turns it into an ∞-category enriched in the monoidal stable ∞-category $D(\Lambda)$ of complexes of Λ-modules (i.e. the localization of the category of chain complexes of Λ-modules by the class of quasi-isomorphisms). In particular, for any objects $\mathcal{F}$ and $\mathcal{G}$ of $D(Sh_h(X, \Lambda))$, morphisms from $\mathcal{F}$ to $\mathcal{G}$ form an object $\mathrm{Hom}(\mathcal{F}, \mathcal{G})$ of $D(\Lambda)$. The appropriate version of the Yoneda Lemma thus reads:

$$\mathrm{Hom}(\Lambda(Y), \mathcal{F}) \cong \mathcal{F}(Y)$$

for any separated X-scheme of finite type Y. In particular, $H^i(Y, \mathcal{F}) = H^i(\mathcal{F}(Y))$ is what the old fashioned literature would call the i-th hypercohomology group of Y with coefficients in $\mathcal{F}$.

---

[2]We refer to [31, 32] in general. However, most of the literature on motives is written using the theory of Quillen model structures. The precise way to translate this language to the one of ∞-categories is discussed in Chapter 7 of [11].

3.2.2.2 A sheaf $\mathcal{F}$ is called $\mathbf{A}^1$-*local* if $\mathcal{F}(Y) \to \mathcal{F}(Y \times \mathbf{A}^1)$ is an equivalence for all Y. A map $f : M \to N$ is an $\mathbf{A}^1$-*equivalence* if for every $\mathbf{A}^1$-local $\mathcal{F}$ the map

$$f^* : \mathrm{Hom}(N, \mathcal{F}) \to \mathrm{Hom}(M, \mathcal{F})$$

is an equivalence.

Define

$$\underline{DM}_h^{eff}(X, \Lambda)$$

to be the localization of $D(Sh_h(X, \Lambda))$ with respect to $\mathbf{A}^1$-equivalences. We have a localization functor $D(Sh_h(X, \Lambda)) \to \underline{DM}_h^{eff}(X, \Lambda)$ with fully faithfull right adjoint whose essential image consists of the $\mathbf{A}^1$-local objects. An explicit description of the right adjoint is by taking the total complex of the bicomplex

$$C_*(\mathcal{F})(Y) = \cdots \to \mathcal{F}(Y \times \Delta_{\mathbf{A}^1}^n) \to \cdots \to \mathcal{F}(Y \times \Delta_{\mathbf{A}^1}^1) \to \mathcal{F}(Y),$$

where $\Delta_{\mathbf{A}^1}^n = Spec(k[x_0, \ldots, x_n]/(x_0 + \cdots + x_n = 1))$. The ∞-category $\underline{DM}_h^{eff}(X, \Lambda)$ comes equipped with a canonical functor

$$\gamma_X : Sch/X \times D(\Lambda) \to \underline{DM}_h^{eff}(X, \Lambda)$$

defined by $\gamma_X(Y, K) = \Lambda(Y) \otimes_\Lambda K$. Furthermore, it is a presentable ∞-category (as a left Bousfield localization of a presentable ∞-category, namely $D(Sh_h(X, \Lambda))$), and thus has small colimits and small limits. For a cocomplete ∞-category C, the category of colimit preserving functors $\underline{DM}_h^{eff}(X, \Lambda) \to C$ is equivalent to the category of functors $F : Sch/X \times D(\Lambda) \to C$ with the following two properties:

- For each X-scheme Y, the functor $F(Y, -) : D(\Lambda) \to C$ commutes with colimits.
- For each complex of Λ-modules K, we have:

  (a) the first projection induces an equivalence $F(Y \times \mathbf{A}^1, K) \cong F(Y, K)$ for any X-scheme Y;

  (b) for any h-hypercovering U of Y, the induced map $\mathrm{colim}_{\Delta^{op}} F(U, K) \to F(Y, K)$ is invertible.

The functor $\underline{DM}_h^{eff}(X, \Lambda) \to C$ associated to such an F is constructed as the left Kan extension of F along γ_X.

There is still an issue. Indeed, let $\infty \in \mathbf{P}^1$ and let us form the following cofiber sequence:

$$\Lambda(X) \xrightarrow{\infty} \Lambda(\mathbf{P}^1) \to \Lambda(1)[2]$$

In order to express Poincaré duality (or, more generally, Verdier duality), we need the cofiber $\Lambda(1)[2]$ above to be $\otimes$-invertible. But it is not so.

Definition 3.2.2.3 An object $A \in C$ is $\otimes$-*invertible* if the functor $A \otimes - : C \to C$ is an equivalence of ∞-categories.

We want to invert a non-invertible object. Let us think about the case of a ring.

$$R[f^{-1}] = \text{colim}(R \xrightarrow{f} R \xrightarrow{f} \cdots)$$

(The colimit is taken within R-modules.) For ∞-categories, we define $C[A^{-1}]$ with a similar colimit formula. Note however that the colimit needs to be taken in the category of presentable ∞-categories (in which the maps are the colimit preserving functors). We get an explicit description of this colimit as follows. For C presentable, $C[A^{-1}]$ can be described as the limit of the diagram

$$\cdots \xrightarrow{Hom(A,-)} C \xrightarrow{Hom(A,-)} C \xrightarrow{Hom(A,-)} C$$

in the ∞-category of ∞-categories (here, $Hom(A, -)$ is the right adjoint of the functor $A \otimes -$). Therefore, an object in $C[A^{-1}]$ is typically a sequence $(M_n, \sigma_n)_{n \geq 0}$ with M_n objects of C and $\sigma_n : M_n \xrightarrow{\sim} Hom(A, M_{n+1})$ equivalences in C. Note that, in the case where A is the circle in the ∞-category of pointed homotopy types, we get exactly the definition of an Ω-spectrum from topology. There is a canonical functor

$$\Sigma^\infty : C \to C[A^{-1}]$$

which is left adjoint to the functor

$$\Omega^\infty : C[A^{-1}] \to C$$

defined as $\Omega^\infty(M) = M_0$ where $M = (M_n, \sigma_n)_{n \geq 0}$ is a sequence as above.

There is still the issue of having a natural symmetric monoidal structure on $C[A^{-1}]$, which is not automatic. However, if the cyclic permutation acts as the identity on $A^{\otimes 3}$ (by permuting the factors) in the homotopy category of C, then there is a unique symmetric monoidal structure on $C[A^{-1}]$ such that the canonical functor $\Sigma^\infty : C \to C[A^{-1}]$ is symmetric monoidal (all these issues are very well explained in Robalo's [39]). Fortunately for us, Voevodsky proved that this extra property holds for $C = \underline{DM}_h^{eff}(X, \Lambda)$ and $A = \Lambda(1)$.

Definition 3.2.2.4 The big category of h-motives is defined as:

$$\underline{DM}_h(X, \Lambda) = \underline{DM}_h^{eff}(X, \Lambda)[\Lambda(1)^{-1}].$$

Remark 3.2.2.5 However, what is important here is the universal property of the stable ∞-category $\underline{DM}_h(X, \Lambda)$; given a cocomplete ∞-category C, together with an equivalence of categories $T : C \to C$ each colimit preserving functor $\varphi : \underline{DM}_h^{eff}(X, \Lambda) \to C$ equipped with an invertible natural transformation $\varphi(M \otimes \Lambda(1)[2]) \cong T(\varphi(M))$ is the composition of a unique colimit preserving functor $\Phi : \underline{DM}_h(X, \Lambda) \to C$ equipped with an invertible natural transformation $\Phi(M \otimes \Sigma^\infty \Lambda(1)[2]) \cong T(\Phi(M))$.

Remark 3.2.2.6 We are very far from having locally constant sheaves here! In classical settings, the Tate object $\Lambda(1)$ is locally constant (more generally, for a smooth and proper map $f : X \to Y$ we expect each cohomology sheaf $R^i f_*(\Lambda)$ to be locally constant). However the special case of the projective line shows that we cannot have such a property motivically: Over field k, the cohomology with coefficients in $\mathbf{Q}$ vanishes in degree >0, while, with coefficients in $\mathbf{Q}(1)$, it is equal to $k^\times \otimes \mathbf{Q}$ in degree 1. Therefore we should ask what is the replacement of locally constant sheaves. This will be dealt with later, when we will explain what are constructible motives.

Definition 3.2.2.7 We have an adjunction

$$\Sigma^\infty : \underline{DM}_h^{eff}(X, \Lambda) \rightleftarrows \underline{DM}_h(X, \Lambda) : \Omega^\infty$$

and we define $M(Y) = \Sigma^\infty \Lambda(Y)$. This is the *motive* of Y over X, with coefficents in Λ.

As we want eventually to do intersection theory, we need Chern classes within motives. Here is how they appear. Consider the morphisms of h-sheaves of groups $\mathbf{Z}(\mathbf{A}^1 - \{0\}) \to \mathbf{G}_m$ on the category Sch/X corresponding to the identity $\mathbf{A}^1 - \{0\} = \mathbf{G}_m$, seen as a map of sheaves of sets. From the pushout diagram

$$\begin{array}{ccc} \mathbf{A}^1 - \{0\} & \longrightarrow & \mathbf{A}^1 \\ \downarrow & & \downarrow \\ \mathbf{A}^1 & \longrightarrow & \mathbf{P}^1 \end{array}$$

and from the identification $\mathbf{Z} \cong \mathbf{Z}(\mathbf{A}^1)$, we get a (split) cofiber sequence

$$\mathbf{Z} \to \mathbf{Z}(\mathbf{A}^1 - \{0\}) \to \mathbf{Z}(1)[1]$$

Since the map $\mathbf{Z}(\mathbf{A}^1 - \{0\}) \to \mathbf{G}_m$ takes $\mathbf{Z}$ to 0, it induces a canonical map $\mathbf{Z}(1)[1] \to \mathbf{G}_m$.

Theorem 3.2.2.8 (Voevodsky) *The map $\mathbf{Z}(1)[1] \to \mathbf{G}_m$ is an equivalence in the effective category $\underline{DM}_h^{eff}(X, \mathbf{Z})$.*

As a result, we get canonical maps:

- h-hypersheafification:

$$Pic(X) = H^1_{Zar}(X, \mathbf{G}_m) \to H^0 \mathrm{Hom}_{D(Sh_h(X,\Lambda))}(\mathbf{Z}, \mathbf{G}_m[1]) \, ;$$

- $\mathbf{A}^1$-localization:

$$H^0 \mathrm{Hom}_{D(Sh_h(X,\Lambda))}(\mathbf{Z}, \mathbf{G}_m[1]) \to H^0 \mathrm{Hom}_{DM_h^{eff}}(\mathbf{Z}, \mathbf{G}_m[1]) \, ;$$

- $\mathbf{P}^1$-stabilization:

$$H^0 \mathrm{Hom}_{DM_h^{eff}(X,\mathbf{Z})}(\mathbf{Z}, \mathbf{G}_m[1]) \to H^0 \mathrm{Hom}_{DM_h(X,\mathbf{Z})}(\mathbf{Z}, \mathbf{Z}(1)[2]) \, .$$

By composition this gives us the first motivic Chern classes of line bundles.

$$c_1 : Pic(X) \to H^2_M(X, \mathbf{Z}(1)) = H^0 \mathrm{Hom}_{DM_h(X,\mathbf{Z})}(\mathbf{Z}, \mathbf{Z}(1)[2])$$

3.2.3 Functoriality

3.2.3.1 Recall that we have an assignment

$$X \mapsto \underline{DM}_h(X, \Lambda).$$

There is a unique symmetric monoidal structure on $\underline{DM}_h(X, \Lambda)$ such that the functor $M : Sch_{/X} \to \underline{DM}_h(X, \Lambda)$ is monoidal. It has the following properties (we write $\Lambda = M(X) \cong \Sigma^\infty(\Lambda)$ and $\Lambda(1) = \Sigma^\infty(\Lambda(1))$):

- $A(1) \cong A \otimes \Lambda(1)$; all functors of interest always commute with the functor $A \mapsto A(1)$.
- $M(Y \times \mathbf{P}^1) \cong M(Y)[2] \oplus M(Y)$.
- $A(n) = A \otimes \Lambda(n)$ is well defined for all $n \in \mathbf{Z}$ (with $\Lambda(n)$ the dual of $\Lambda(-n)$ for $n < 0$ and $\Lambda(0) = \Lambda$, $\Lambda(n + 1) \cong \Lambda(n)(1)$ for $n \geq 0$).
- There is an internal Hom functor *Hom*.

For a morphism $f : X \to Y$ we have $f^* : \underline{DM}_h(Y, \Lambda) \to \underline{DM}_h(X, \Lambda)$ which preserves colimits and thus has right adjoint $f_* : \underline{DM}_h(X, \Lambda) \to \underline{DM}_h(Y, \Lambda)$. No property of f is required for that. We construct first the functor

$$f^* : \underline{DM}_h^{eff}(Y, \Lambda) \to \underline{DM}_h^{eff}(X, \Lambda)$$

as the unique colimit preserving functor which fits in the commutative diagram

$$Sch/Y \times D(\Lambda) \xrightarrow{f^* \times 1_{D(\Lambda)}} Sch/X \times D(\Lambda)$$

$$\downarrow \qquad\qquad\qquad \downarrow$$

$$\underline{DM}_h^{eff}(Y, \Lambda) \xrightarrow{\quad f^* \quad} \underline{DM}_h^{eff}(X, \Lambda)$$

(in which the vertical functors are the canonical ones $(U, C) \mapsto M(U) \otimes_\Lambda C$), and observe that it has a natural structure of symmetric monoidal functor. There is thus a unique symmetric monoidal pull-back functor f^* defined on $\underline{DM}_h$ so that the following squares commutes.

$$\underline{DM}_h^{eff}(Y, \Lambda) \xrightarrow{\quad f^* \quad} \underline{DM}_h^{eff}(X, \Lambda)$$

$$\downarrow{\Sigma^\infty} \qquad\qquad\qquad \downarrow{\Sigma^\infty}$$

$$\underline{DM}_h(Y, \Lambda) \xrightarrow{\quad f^* \quad} \underline{DM}_h(X, \Lambda)$$

If moreover f is separated and of finite type then the pull-back functor f^* has a left adjoint functor $f_\sharp : \underline{DM}_h(X, \Lambda) \to \underline{DM}_h(Y, \Lambda)$ which preserves colimits, and is essentially determined by the property that $f_\sharp M(U) = M(U)$ for any separated X-scheme of finite type U via universal properties as above. For example $f_\sharp(\Lambda) = M(X)$. We have a projection formula (proved by observing that the formula holds in the category of schemes and then extending by colimits)

$$f_\sharp(A \otimes f^*(B)) \xrightarrow{\simeq} f_\sharp A \otimes B.$$

Exercise 3.2.3.2 Show that, for any Cartesian square of noetherian schemes

$$\begin{array}{ccc} X' & \xrightarrow{u} & X \\ \downarrow{f'} & & \downarrow{f} \\ Y' & \xrightarrow{v} & Y \end{array}$$

and for any M in $\underline{DM}_h(X, \Lambda)$, if v is separated of finite type, then the canonical map

$$v^* f_*(M) \to f'_* u^*(M)$$

is invertible.

The base change formula above is too much: we want this to hold only for f proper of v smooth, because, otherwise, we will not have any good notion of support of a motive. This is why we have to restrict ourselves to a subcatefgory of $\underline{DM}_h(X, \Lambda)$, on which the support will be well defined.

Definition 3.2.3.3 Let $DM_h(X, \Lambda)$ be the smallest full subcategory of $\underline{DM}_h(X, \Lambda)$ closed under small colimits, containing objects of the form $M(U)(n)[i]$ for $U \to X$ smooth and $i, n \in \mathbf{Z}$.

Remark 3.2.3.4 The ∞-category $DM_h(X, \Lambda)$ is stable and presentable, essentially by construction. It is also stable under the operator $M \mapsto M(n)$ for all $n \in \mathbf{Z}$.

Theorem 3.2.3.5 (Localization Property) *Take $i : Z \to X$ to be a closed emdebbing with open complement $j : U \to X$ and let $M \in DM_h(X, \Lambda)$. Then we have a canonical cofiber sequence (in which the maps are the co-unit and unit of appropriate adjunctions):*

$$j_\sharp j^* M \to M \to i_* i^* M$$

Idea of the proof: the functors $j_\sharp$, j^*, i_* and i^* commute with colimits. Therefore, it is sufficient to prove the case where $M = M(U)$ with U/X smooth. We conclude by an argument due to Morel and Voevodsky, using Nisnevich excision as well as the fact, locally for the Zariski topology, U is étale on $\mathbf{A}^n \times X$. Then, using Nisnevich excision, we reduce to the vase where $U = \mathbf{A}^n \times X$, in which case we can provide explicit $\mathbf{A}^1$-homotopies.

Exercise 3.2.3.6 Show that $j_\sharp j^* M \to M \to i_* i^* M$ is not a cofiber sequence in $\underline{DM}_h(X, \Lambda)$ for an arbitrary object M.

The functor f^* restricts to a functor on DM_h, and also for $f_\sharp$ if f is smooth. Moreover, DM_h is closed under tensor product. If $i : Z \to X$ is a closed immersion, than by the cofiber sequence above we see that the functor i_* sends $DM_h(Z, \Lambda)$ to $DM_h(X, \Lambda)$.

Remark 3.2.3.7 By presentability, the inclusion $DM_h(X, \Lambda) \overset{i}{\to} \underline{DM}_h(X, \Lambda)$ has right adjoint ρ.

For $f : X \to Y$ we define

$$f_* : DM_h(X, \Lambda) \to DM_h(Y, \Lambda)$$

by

$$f_* M = \rho f_* i(M)$$

We can use this to describe the internal Hom as well:

$$Hom(A, B) = \sigma Hom(i(A), i(B)) \,.$$

Proposition 3.2.3.8 *For any embedding $i : Z \to X$ the functors $i_*, i_\sharp$ are both fully faithful.*

Using this and some abstract nonsense we get that i_* has a right adjoint $i^!$ and there are canonical fiber sequences

$$i_* i^! M \to M \to j_* j^* M$$

We also have a smooth base change formula and a proper base change formula:

Theorem 3.2.3.9 (Ayoub, Cisinski-Déglise) *For any Cartesian square of noetherian schemes*

$$
\begin{array}{ccc}
X' & \xrightarrow{\ u\ } & X \\
\downarrow{f'} & & \downarrow{f} \\
Y' & \xrightarrow{\ v\ } & Y
\end{array}
$$

and for any M in $DM_h(X, \Lambda)$, if either v is separated smooth of finite type, or if f is proper, then the canonical map

$$v^* f_*(M) \to f'_* u^*(M)$$

is invertible in $DM_h(X, \Lambda)$.

The proof follows from Ayoub's axiomatic approach [4], under the additional assumption that all the maps are quasi-projective. The general case may be found in [14, Theorem 2.4.12].

Definition 3.2.3.10 (Deligne) Let $f : X \to Y$ be separated of finite type, or equivalently, by Nagata's theorem, assume that there is a relative compactification which is a factorization of f as

$$X \xrightarrow{\ j\ } \bar{X} \xrightarrow{\ p\ } Y,$$

where j is an open embedding and p is proper. Then we define

$$f_! = p_* j_\sharp$$

Here are the main properties we will use (see [14]):

- The functor $f_!$ admits a right adjoint $f^!$ (because it commutes with colimits).
- There is a comparison map $f_! \to f_*$ constructed as follows. There is a map $j_\sharp \to j_*$ which corresponds by transposition to the inverse of the isomorphism from $j^* j_*$ to the identity due to the fully faithfulness of j_*. Therefore we have a map $f_! = p_* j_\sharp \to p_* j_* \cong f_*$.
- Using the proper base change formula, we can prove that push-forwards with compact support are well defined: in particular, the functor $f_!$ does not depend

on the choice of the compactification of f up to isomorphism. Furthermore, if f and g are composable, there is a coherent isomorphism $f_! g_! \cong (fg)_!$.

The proof of the proper base change formula relies heavily on the following property.

Theorem 3.2.3.11 (Relative Purity) *If $f : X \to Y$ is smooth and separated of finite type, then*

$$f^!(M) \cong f^*(M)(d)[2d]$$

where $d = dim(X/Y)$.

The first appearance of this kind of result in a motivic context (i.e. in stable homotopy category of schemes) was in a preprint of Oliver Röndigs [40]. As a matter of facts, the proof of relative purity can be made with a great level of generality, as in Ayoub's thesis [4], where we see that the only inputs are the localization theorem and $\mathbf{A}^1$-homotopy invariance. However, in our situation (where Chern classes are available), the proof can be dramatically simplified (see the proof [13, Theorem 4.2.6], which can easily be adapted to the context of h-sheaves). A very neat and robust proof (in equivariant stable homotopy category of schemes, but which may be seen in any context with the six operations) may be found in Hoyois' paper [24].

Remark 3.2.3.12 For a vector bundle $E \to X$ of rank r, we can define its *Thom space* $Th(E)$ by the cofiber sequence

$$\Lambda(E - 0) \to \Lambda(E) \to Th(E)$$

(where $E - 0$ is the complement of the zero section). Using motivic Chern classes, we can construct the Thom isomorphism

$$Th(E) \cong \Lambda(r)[2r].$$

What is really canonical and conceptually right is

$$f^!(M) \cong f^*(M) \otimes Th(T_f).$$

We refer to Ayoub's work for more details. From this we can deduce a formula relating $f_!$ and $f_\sharp$ when f is smooth. By transposition, relative putity takes the following form.

Corollary 3.2.3.13 *If $f : X \to Y$ is smooth and separated of finite type then*

$$f_\sharp(M) \cong f_!(M)(d)[2d].$$

Finally, we also need the Projection Formula (see [14, Theorem 2.2.14]):

Proposition 3.2.3.14 *If $f : X \to Y$ is separated of finite type then, for any A in $DM_h(X, \Lambda)$ and any B in $DM_h(Y, \Lambda)$, there is a canonical isomorphism:*

$$f_!(A \otimes f^*B) \cong f_!(A) \otimes B.$$

Exercise 3.2.3.15

- Let $f : X \to Y$, then $f_*Hom(f^*M, N) \cong Hom(M, f_*N)$.
- For f separated of finite type we have $Hom(f_!M, N) \cong f_*Hom(M, f^!N)$.
- For f as above, $f^!Hom(M, N) \cong Hom(f^*M, f^!N)$.
- For f smooth, $f^*Hom(M, N) \cong Hom(f^*M, f^*N)$.

A reformulation of the proper base change formula is the following.

Theorem 3.2.3.16 *For any pull-back square of noetherian schemes*

$$
\begin{array}{ccc}
X' & \xrightarrow{\ u\ } & X \\
\downarrow{\scriptstyle f'} & & \downarrow{\scriptstyle f} \\
Y' & \xrightarrow{\ v\ } & Y
\end{array}
$$

with f is separated of finite type, we have $v^* f_! \cong f'_! u^*$ and $f^! v_* \cong u_*(f')^!$.

Remark 3.2.3.17 Given a morphism of rings of coefficients $\Lambda \to \Lambda'$, there is an obvious change of coefficients functor

$$DM_h(X, \Lambda) \to DM_h(X, \Lambda'), \quad M \mapsto \Lambda' \otimes_\Lambda M$$

which is symmetric monoidal and commutes with the four operations f^*, f_*, $f^!$ and $f_!$ whenever they are defined. Moreover, one can show that an object M in $DM_h(X, \mathbf{Z})$ is null if and only if $\mathbf{Q} \otimes M \cong 0$ and $\mathbf{Z}/p\mathbf{Z} \otimes M \cong 0$ for any prime number p; see [13, Prop. 5.4.12]. Fortunately, $DM_h(X, \Lambda)$ may be understood in more tractable terms whenever $\Lambda = \mathbf{Q}$ of Λ is finite, as we will see in the next section.

3.2.4   Representability Theorems

3.2.4.1 We define *étale motivic cohomology*[3] of X with coefficients in Λ as

$$H^i_M(X, \Lambda(n)) = H^i(\mathrm{Hom}_{DM_h(X,\Lambda)}(\Lambda, \Lambda(n)))$$

for all $i, n \in \mathbf{Z}$.

---

[3] Also known as Lichtenbaum cohomology.

Theorem 3.2.4.2 (Suslin-Voevodsky, Cisinski-Déglise) *For any noetherian scheme of finite dimension X,*

$$H_M^i(X, \mathbf{Q}(n)) \cong (K H_{2n-i}(X) \otimes \mathbf{Q})^{(n)}$$

where $K H$ is the homotopy invariant K-theory of Weibel and the "(n)" stands for the fact that we take the intersection of the k^n-eigen-spaces of the Adams operations ψ_k for all k. For X regular, we simply have $H_M^i(X, \mathbf{Q}(n)) \cong (K_{2n-i}(X) \otimes \mathbf{Q})^{(n)}$. In particular, for X regular and $n \in \mathbf{Z}$, we have:

$$CH^n(X) \otimes \mathbf{Q} \cong H_M^{2n}(X, \mathbf{Q}(n)).$$

The case where X is separated and smooth of finite type over a field is due to Suslin and Voevodsky (puting together the results of [43] and of [49]). The general case follows from [13, Theorem 5.2.2], using the representability theorem of $K H$ announced in [48] and proved in [10]. More generally, one may recover motivically $\mathbf{Q}$-linear Chow groups of possibly singular schemes as well as Bloch's higher Chow groups as follows.

Theorem 3.2.4.3 (Motivic Cycle Class) *Let $f : X \to Spec(k)$ be separated of finite type. Then*

$$H^0(\mathrm{Hom}_{DM_h(X,\mathbf{Q})}(\mathbf{Q}(n)[2n], f^!\mathbf{Q})) \cong CH_n(X) \otimes \mathbf{Q}$$

and, if X is equidimensional of dimension d, then

$$H^0(\mathrm{Hom}_{DM_h(X,\mathbf{Q})}(\mathbf{Q}(n)[i], f^!\mathbf{Q})) \cong CH^{d-n}(X, i - 2n) \otimes \mathbf{Q}.$$

This follows from [12, Corollaries 8.12 and 8.13, Remark 9.7]. These representability result may be used to see how classical Grothendieck motives of smooth projective varieties over a field k may be seen in this picture: they form the full subcategory of $DM_h(\mathrm{Spec}(k), \mathbf{Q})$ whose objects are the direct factors of motives of the form $M(U)(n)$ with U smooth and projective over k and $n \in \mathbf{Z}$). The following statement is known as *rigidity theorem*

Theorem 3.2.4.4 (Suslin-Voevodsky, Cisinski-Déglise) *Given a locally noetherian scheme X, there is a canonical equivalence of ∞-categories*

$$DM_h(X, \Lambda) \cong D(Sh(X_{et}, \Lambda))$$

for Λ of positive invertible characteristic on X, compatible with 6-operations. In particular

$$H_M^i(X, \Lambda(j)) \cong H_{et}^i(X, \mu_n^{\otimes j} \otimes \Lambda).$$

The case where X is the spectrum of a field is essentially contained in the work of Suslin and Voevodsky [43]. See [13, Corollary 5.5.4] for the general case. We should mention that the equivalence of categories above is easy to construct. The main observation is Voevodsky's Theorem 3.2.2.8, together with the Kummer short exact sequence induced by $t \mapsto t^n$

$$0 \to \mu_n \to \mathbf{G}_m \to \mathbf{G}_m \to 0$$

(where μ_n is the sheaf of n-th roots of unity), from which follows the identification $\Lambda(1) \cong \mu_n \otimes_{\mathbf{Z}/n\mathbf{Z}} \Lambda$, where n is the characteristic of Λ. In particular, $\Lambda(1)$ is already $\otimes$-invertible, which implies (by inspection of universal properties) that

$$\underline{DM}_h^{eff}(X, \Lambda) \cong \underline{DM}_h(X, \Lambda).$$

On the other hand, $\underline{DM}_h^{eff}(X, \Lambda)$ is a full subcategory of the derived category of h-sheaves of Λ-modules. The comparison functor from $\underline{DM}_h^{eff}(X, \Lambda)$ to $D(Sh(X_{et}, \Lambda))$ is simply the restriction functor. The precise formulation of the previous theorem is that the composition

$$DM_h(X, \Lambda) \subset \underline{DM}_h(X, \Lambda) \cong \underline{DM}_h^{eff}(X, \Lambda) \to D(Sh(X_{et}, \Lambda))$$

is an equivalence of ∞-categories.

Remark 3.2.4.5 If $char(\Lambda) = p^i$ then one proves that $DM_h(X, \Lambda) \cong DM_h(X[\frac{1}{p}], \Lambda)$ (using the Artin-Schreier short exact sequence together with the localization property) so that we can assume that the ring of functions on X always has the characteristic of Λ invertible in it; see [13].

Remark 3.2.4.6 One can have access to $H_M^i(X, \mathbf{Z}(n))$ via the coniveau spectral sequence whose E_1 term is computed as Cousin complex, and thus gives rise to a nice and rather explicit theory of residues; see [13, (7.1.6.a) and Prop. 7.1.10].

3.3 Finiteness and Euler Characteristic

3.3.1 Locally Constructible Motives

3.3.1.1 Recall that an object X in a tensor category C is *dualizable* (we also say *rigid*) if there exists $Y \in C$ such that $X \otimes -$ is left adjoint to $Y \otimes -$. This provides an isomorphism $Y \cong Hom(X, 1_C)$. In other words $Y \otimes a \cong Hom(X, a)$. This way, we get the evaluation map $\epsilon : Y \otimes X \to 1_C$ and as well as the co-evaluation map $\eta : 1_C \to X \otimes Y$. This exhibits the adjunction between the tensors. In particular, composing ϵ and η appropriately tensored by X or Y gives the identity:

$$1_X : X \to X \otimes Y \otimes X \to X \quad \text{and} \quad 1_Y : Y \to Y \otimes X \otimes Y \to Y.$$

Remark 3.3.1.2 If $F : C \to D$ is a monoidal functor, if $x \in C$ dualizable then so is $F(x)$, and $F(x^\wedge) \cong F(x)^\wedge$. Furthermore, F also preserve internal *Hom* from x, since $Hom(x, y) \cong x^\wedge \otimes y$ for all y.

Remark 3.3.1.3 If $C \in D(Sh_{et}(X, \Lambda))$ then it is dualizable if and only if it is locally constant with perfect fibers; see [13, Remark 6.3.27]. That means that C is dualizable if and only if the following condition holds: there is a surjective étale map $u : X' \to X$ together with a perfect complex of Λ-modules $K \in Perf(\Lambda)$ (i.e. complex of Λ-modules K which is quasi-isomorphic to a bounded complex of projective Λ-modules of finite type), and an isomorphism $K_{X'} \cong u^*(C)$ in $D(Sh_{et}(X', \Lambda))$, where $K_{X'}$ is the constant sheaf on X' associated to K.

3.3.1.4 Suppose $1/n \in \mathcal{O}_X$, $n = char(\Lambda) > 0$. Then $DM_h(X, \Lambda) \cong D(Sh_{et}(X, \Lambda))$. Inside it, we have the subcategory $D^b_{ctf}(X_{et}, \Lambda)$ of constructible sheaves finite tor-dimension. If there is d such that $cd(k(x)) \le d$ for every point x of X, then it is simply the subcategory of compact objects. In general, this subcategory $D^b_{ctf}(X_{et}, \Lambda)$ is important because it is closed under the six operations. We look for correspondent in motives with arbitrary ring of coefficients Λ. We can characterise those étale sheaves by

$$\{C \in D(Sh_{et}(X, \Lambda)) \mid \exists \text{ stratification } X_i : C_{|X_i} \text{ locally constant with perfect fibers}\}$$

Namely, an object C of $D(Sh_{et}(X, \Lambda))$ is constructible of finite tor-dimension if and only if there exists a finite stratification of X by locally closed subschemes X_i together with $\phi_i : U_i \to X_i$ étale surjective for each i, and there is $K_i \in Perf(\Lambda)$ (compact objects in the derived category of Λ-modules), and an isomorphism $\phi_i^*(C_{|X_i}) \cong (K_i)_{U_i}$ in the derived category of sheaves of Λ-modules on the small étale site of U_i; see [13, Remark 6.3.27].

Exercise 3.3.1.5 (Poincaré Duality) Let $f : X \to Y$ be smooth and proper of relative dimension d. Then, if $M \in DM_h(X, \Lambda)$ is dualizable, so is $f_*(M)$ and

$$f_*(M)^\wedge \cong f_*(M^\wedge)(-d)[-2d]$$

with $M^\wedge = Hom(M, \Lambda)$ the dual of M.

Definition 3.3.1.6 The ∞-category $DM_{h,c}(X, \Lambda)$ of *constructible* Λ-*linear* étale motives over X is the smallest thick subcategory (closed under shifts, finite colimits and retracts) containing $f_\sharp(\Lambda)(n)$ for any $f : U \to X$ smooth and every $n \in \mathbf{Z}$.

The following proposition is an easy consequence of relative purity and of the proper base change formula.

Proposition 3.3.1.7 *The ∞-category $DM_{h,c}(X, \Lambda)$ is equal to each of the following subcategories of $DM_h(X, \Lambda)$:*

- *The smallest thick subcategory containing $f_*(\Lambda)(n)$ for $f : U \to X$ proper and $n \in \mathbf{Z}$.*

- *The smallest thick subcategory containing $f_!(\Lambda)(n)$ for $f : U \to X$ separated of finite type and $n \in \mathbf{Z}$.*

Theorem 3.3.1.8 (Absolute Purity) *If $i : Z \to X$ is a closed emmersion and assume that both X, Z are regular. Let $c = codim(Z, X)$. Then there is a canonical isomorphism*

$$i^!(\Lambda_X) \cong \Lambda_Z(-c)[-2c].$$

See [13, Theorem 5.6.2]

Remark 3.3.1.9 Modulo the rigidity Theorem 3.2.4.4, the proof for the case of finite coefficients is due to Gabber and was known for a while, with two different proofs [20, 26] (although, in characteristic zero, this goes back to Artin in SGA 4). After formal reductions using deformation to the normal cone, one sees that, in order to prove the absolute purity theorem above, it is then sufficient to consider the case where $\Lambda = \mathbf{Q}$. The idea is then that Quillen's localization fiber sequence

$$
\begin{array}{ccccc}
K(Z) & \longrightarrow & K(X) & \longrightarrow & K(X-Z) \\
\downarrow{\scriptstyle\wr} & & \downarrow{\scriptstyle\wr} & & \downarrow{\scriptstyle\wr} \\
K(Coh(Z)) & \longrightarrow & K(Coh(X)) & \longrightarrow & K(Coh(X-Z))
\end{array}
$$

induces a long exact sequence which we may tensor with $\mathbf{Q}$, and Absolute purity is then proved using the representability theorem of K-theory in the motivic stable homotopy category together with a variation on the Adams-Riemann-Roch theorem.

We recall that a locally noetherian scheme X is *quasi-excellent* if the following two conditions are verified:

1. For any point $x \in X$, the completion map $\mathcal{O}_{X,x} \to \hat{\mathcal{O}}_{X,x}$ is regular (i.e., for any field extension K of the residue field $\kappa(x)$, the noetherian ring $K \otimes_{\kappa(x)} \hat{\mathcal{O}}_{X,x}$ is regular).
2. For any scheme of finite presentation Y over X, there is a regular dense open subscheme $U \subset Y$.

A locally noetherian scheme is *excellent* if it is quasi-excellent and universally catenary. In practice, what needs to be known is that any scheme of finite type over a quasi-excellent scheme is quasi-excellent, and $Spec(R)$ is excellent whenever R is either a field or the ring of integers of a number field (note also that noetherian complete local rings are excellent).

Theorem 3.3.1.10 (de Jong-Gabber [26]) *Any quasi-excellent scheme is regular locally for the h-topology. In other words, for any quasi-excellent scheme X, there exists an h-covering $\{X_i \to X\}_i$ with each X_i regular. Furthermore, locally for the h-topology any nowhere dense closed subscheme of X is either empty of a divisor with normal crossings: given any nowhere dense closed subscheme $Z \subset X$, we may* *choose the covering above such that the pullback of Z in each X_i is either empty or a divisor with normal crossings. Even better, given a prime ℓ invertible in $\mathcal{O}_X$, we may always choose h-coverings $\{X_i \to X\}_i$ as above such that, for each point $x \in X$, there exists an i and there exists $x_i \in X_i$ such that $p_i(x_i) = x$ and such that $[k(x_i) : k(x)]$ is prime to ℓ.*

Remark 3.3.1.11 One can show that the category $DM_{h,c}(X, \Lambda)$ is preserved by the 6 operations. However, there is a drawback: unless we make finite cohomological dimension assumptions, the category $DM_{h,c}$ in not always a sheaf for the étale topology! Here is its étale sheafification (which can be proved to be a sheaf of ∞-categories for the h-topology).

Definition 3.3.1.12 A motivic sheaf M is in $DM_h(X, \Lambda)$ is *locally constructible* if there is an étale surjection $f : U \to X$ such that $f^*M \in DM_{h,c}(X, \Lambda)$.

Denote the full subcategory of locally constructible motives by $DM_{h,lc}(X, \Lambda)$.

Remark 3.3.1.13 If $\mathbf{Q} \subset \Lambda$, then $DM_{h,c}(X, \Lambda) = DM_{h,lc}(X, \Lambda)$ simply is the full subcategory of compact objects in $DM_h(X, \Lambda)$; see [13, Prop. 6.3.3].

Theorem 3.3.1.14 (Cisinski-Déglise) *The equivalence $DM_h(X, \Lambda) \cong D(X_{et}, \Lambda)$ restricts to an equivalence of ∞-categories*

$$DM_{h,lc}(X, \Lambda) \cong D^b_{ctf}(X, \Lambda)$$

whenever Λ is noetherian of positive characteristic n, with $\frac{1}{n} \in \mathcal{O}_X$.

See [13, Theorem 6.3.11].

For any morphism of noetherian schemes $f : X \to Y$, the functor f^* sends locally constructible h-motives to locally constructible h-motives, and, in the case where f is separated of finite type, so does the functor $f_!$. The theorem of de Jong-Gabber above, together with Absolute Purity, are the main ingredients in the proof of the following finiteness theorem.

Theorem 3.3.1.15 (Cisinski-Déglise) *The six operations preserve locally constructible h-motives, at least when restricted to separated morphisms of finite type between quasi-excellent noetherian schemes of finite dimension:*

1. *for any such scheme X and any locally constructible h-motives M and N over X, the h-motives $M \otimes N$ and $Hom(M, N)$ are locally constructible;*
2. *for any morphism of finite type $f : X \to Y$ between quasi-excellent noetherian schemes of finite dimension, the four functors f^*, f_*, $f_!$, and $f^!$ preserve the property of being locally constructible.*

See [13, Corollary 6.3.15].

Theorem 3.3.1.16 (Cisinski-Déglise) *Let X be a noetherian scheme of finite dimension, and M an object of $DM_h(X, \Lambda)$.*

1. *If M is dualizable, then it is locally constructible.*

2. *If there exists a closed immersion* $i : Z \to X$ *with open complement* $j : U \to X$ *such that* $i^*(M)$ *and* $j^*(M)$ *are locally constructible, then* M *is locally constructible.*
3. *If* M *is locally constructible over* X, *then there exists a dense open immersion* $j : U \to X$ *such that* $j^*(M)$ *is dualizable in* $DM_{h,lc}(U)$.

This is a reformulation of (part of) [13, Theorem 6.3.26].

Remark 3.3.1.17 In particular, an object M of $DM_h(X, \Lambda)$ is constructible if and only if there exists a finite stratification of X by locally closed subschemes X_i such that each restriction $M_{|X_i}$ is dualizable in $DM_h(X_i, \Lambda)$. This may be seen as an independence of ℓ result. Indeed, as we will recall below, there are ℓ-adic realization functors and they commute with the six functors. In particular, for each appropriate prime number ℓ, the ℓ-adic realization $R_\ell(M)$ is a constructible ℓ-adic sheaf: each restriction $R_\ell(M)_{|X_i}$ is smooth (in the language of SGA 4, 'localement constant tordu'),[4] where the X_i form a stratification of X which is given independently of ℓ. Furthermore, if we apply any of the six operations to $R_\ell(M)$ in the ℓ-adic context, then we obtain an object of the from $R_\ell(N)$ for some locally constructible motive N, and thus a stratification as above relatively to $R_\ell(N)$ which does not depend on ℓ.

3.3.2   Integrality of Traces and Rationality of ζ-Functions

3.3.2.1 For x a dualizable object in a tensor category C with unit object $\mathbf{1}$, we can from the trace of an endomorphism. Indeed the trace of $f : x \to x$ is the map $Tr(f) : \mathbf{1} \to \mathbf{1}$ defined as the composite bellow.

$$\mathbf{1} \xrightarrow{\text{unit}} Hom(x, x) \cong x^\wedge \otimes x \xrightarrow{1 \otimes f} x^\wedge \otimes x \xrightarrow{\text{evaluation}} \mathbf{1}$$

If a functor $\Phi : C \to D$ is symmetric monoidal, then the induced map

$$\Phi : \mathrm{Hom}_C(x, x) \to \mathrm{Hom}_C(\mathbf{1}, \mathbf{1})$$

preserves the formation of traces: $\Phi(Tr(f)) = Tr(\Phi(f))$.

---

[4]It is standard terminology to call such ℓ-adic sheaves 'lisses'. This comes from Deligne's work, which is written in French. I prefer to translate into 'smooth' because this is what we do for morphisms of schemes. The reason is that this terminology comes from the fact that there are transersality conditions one can define between (motivic or ℓ-adic) sheaves and morphisms of schemes, and that a basic intuition about smoothness is that a smooth object is transverse to anything: indeed, a smooth sheaf is transverse to any morphism, while any sheaf is transverse to a smooth morphism. This why I think it is better to use the same word to express the smoothness of both sheaves and morphisms of schemes.

3.3.2.2 If $M \in DM_{h,lc}(Spec(k), \Lambda)$ for k a field (see Definition 3.3.1.12), then M is dualizable. Furthermore, the unit is Λ and we can compute

$$H^0 \mathrm{Hom}_{DM_{h,lc}(Spec(k),\Lambda)}(\Lambda, \Lambda) = \Lambda \otimes \mathbf{Z}[1/p]$$

where p is the exponent characteristic of k (i.e. $p = char(k)$ if $char(k) > 0$ or $p = 1$ else). For $f : M \to M$ any map in $DM_{h,lc}(Spec(k), \mathbf{Z})$, we thus have its trace

$$Tr(f) \in \mathbf{Z}[1/p].$$

The *Euler characteristic* of a dualizable object M of $DM_h(Spec(k), \mathbf{Z})$ is defined as the trace of its identity:

$$\chi(M) = Tr(1_M).$$

For separated k-scheme of finite type X, we define in particular

$$\chi_c(X) = \chi(a_! \mathbf{Z})$$

with $a : X \to Spec(k)$ the structural map.

3.3.2.3 Let X be a noetherian scheme and ℓ a prime number. Let $\mathbf{Z}_{(\ell)}$ be the localization of $\mathbf{Z}$ at the prime ideal (ℓ). We may identify $DM_h(X, \mathbf{Q})$ as the full subcategory of $DM_h(X, \mathbf{Z}_{(\ell)})$ whose objects are the motives M such that $M/\ell M \cong 0$, where $M/\ell M \cong \mathbf{Z}/\ell\mathbf{Z} \otimes M$ is defined via the following cofiber sequence:

$$M \xrightarrow{\ell} M \to M/\ell M.$$

We define

$$\hat{D}(X, \mathbf{Z}_\ell) = DM_h(X, \mathbf{Z}_{(\ell)})/DM_h(X, \mathbf{Q}).$$

In other words, $\hat{D}(X, \mathbf{Z}_\ell)$ is the localization (in the sense of ∞-categories) of $DM_h(X, \mathbf{Z}_{(\ell)})$ by the maps $f : M \to N$ whose cofiber is uniquely ℓ-divisible (i.e. lies in the subcategory $DM_h(X, \mathbf{Q})$). One can show that, if $\frac{1}{\ell} \in \mathcal{O}_X$, the homotopy category of $\hat{D}(X, \mathbf{Z}_\ell)$ is Ekedahl's derived category of ℓ-adic sheaves on the small étale site of X. In fact, as explained in [13, Prop. 7.2.21] (although in the language of model categories), the rigidity Theorem 3.2.4.4 may be interpreted as an equivalence of ∞-categories of the form:

$$\hat{D}(X, \mathbf{Z}_\ell) \cong \lim_n D(X_{et}, \mathbf{Z}/\ell^n \mathbf{Z})$$

(here, the limit is taken in the ∞-categories of ∞-categories). We thus have a canonical ℓ-*adic realization functor*

$$R_\ell : DM_h(X, \mathbf{Z}) \to \lim_n D(X_{et}, \mathbf{Z}/\ell^n\mathbf{Z})$$

which sends a motive M to $M \otimes \mathbf{Z}_{(\ell)}$, seen in the Verdier quotient $\hat{D}(X, \mathbf{Z}_\ell)$. We observe that there is a unique way to define the six operations on $\hat{D}(X, \mathbf{Z}_\ell)$ in such a way that the ℓ-adic realization functor commutes with them. In particular, there is a symmetric monoidal structure on $\hat{D}(X, \mathbf{Z}_\ell)$.

Classically, one defines $D_c^b(X_{et}, \mathbf{Z}_\ell)$ as the full subcategory of $\lim_n D(X_{et}, \mathbf{Z}/\ell^n\mathbf{Z})$ whose objects are the ℓ-adic systems $(\mathcal{F}_n)$ such that each $\mathcal{F}_n$ belongs to the subcategory $D_{ctf}^b(X_{et}, \mathbf{Z}/\ell^n\mathbf{Z})$ (see 3.3.1.4). Furthermore, an ℓ-adic system $(\mathcal{F}_n)$ is dualizable is and only if $\mathcal{F}_1$ is dualizable in $D_{ctf}^b(X_{et}, \mathbf{Z}/\ell\mathbf{Z})$: this is due to the fact, that, by definition, the canonical functor

$$\hat{D}(X, \mathbf{Z}_\ell) \to D(X_{et}, \mathbf{Z}/\ell\mathbf{Z})$$

is symmetric monoidal, conservative, and commutes with the formation of internal Hom's. In other words, $D_c^b(X_{et}, \mathbf{Z}_\ell)$ may be identified with the full subcategory of $\hat{D}(X, \mathbf{Z}_\ell)$ whose objects are those $\mathcal{F}$ such that there exists a finite stratification by locally closed subschemes $X_i \subset X$ such that each restriction $\mathcal{F}_{|X_i}$ is dualizable in $\hat{D}(X_i, \mathbf{Z}_\ell)$. We thus have a canonical equivalence of ∞-categories:

$$D_c^b(X_{et}, \mathbf{Z}_\ell) \cong \lim_n D_{ctf}^b(X_{et}, \mathbf{Z}/\ell^n\mathbf{Z}) .$$

This implies right away that the six operations restrict to $D_c^b(X_{et}, \mathbf{Z}_\ell)$ (if we consider quasi-excellent schemes only), and, by vitue of Theorem 3.3.1.14 that we have an ℓ-adic realization functor

$$R_\ell : DM_{h,lc}(X, \mathbf{Z}) \to D_c^b(X_{et}, \mathbf{Z}_\ell)$$

which commute with the six operations. For a scheme X with structural map $a : X \to \mathrm{Spec}(k)$ separated and of finite type, the motive of X is $M(X) = a_! a^!(\mathbf{Z})$. This is a dualizable object with dual $a_*(\mathbf{Z})$. Hence

$$R_\ell(M(X)^\wedge(n)) = a_*(\mathbf{Z}_\ell(n)) = R\Gamma(X_{et}, \mathbf{Z}_\ell(n))$$

is a dualizable object in $D_c^b(\mathrm{Spec}(k)_{et}, \mathbf{Z}_\ell)$. For k separably closed, the latter category simply is the bounded derived category of $\mathbf{Z}_\ell$-modules of finite type, and this proves in particular that ℓ-adic cohomology

$$H_{et}^i(X, \mathbf{Z}_\ell(n)) = H^i(R_\ell(M(X)^\wedge(n)))$$

is of finite type as a $\mathbf{Z}_\ell$-module for all i (and trivial for all but finitely many i's). Similarly, the ℓ-adic realization of $a_!(\mathbf{Z})$ gives ℓ-adic cohomology with compact support

$$H^i_{et,c}(X, \mathbf{Z}_\ell(n)) = H^i(R_\ell(a_!(\mathbf{Z})(n))) .$$

3.3.2.4 In particular, for any field k of characteristic prime to ℓ, we have a symmetric monoidal functor

$$R_\ell : DM_{h,lc}(k, \mathbf{Z}) \to D^b_c(k, \mathbf{Z}_\ell)$$

inducing the map of rings

$$R_\ell : \mathbf{Z}[1/p] \cong H^0\mathrm{Hom}_{DM_{h,lc}(k,\mathbf{Z})}(\mathbf{Z}, \mathbf{Z}) \to H^0\mathrm{Hom}_{D^b_c(k,\mathbf{Z}_\ell)}(\mathbf{Z}_\ell, \mathbf{Z}_\ell) \cong \mathbf{Z}_\ell .$$

Therefore, for an endomorphism $f : M \to M$ we have $Tr(f) \in \mathbf{Z}[1/p]$ sent to the ℓ-adic number $Tr(R_\ell(f)) \in \mathbf{Z}_\ell$. We thus get:

Corollary 3.3.2.5 *The ℓ-adic trace $Tr(R_\ell(f)) \in \mathbf{Z}[1/p]$ and is independent of ℓ.*

Remark 3.3.2.6 If k is separably closed, then $D^b_c(k, \mathbf{Z}_\ell)$ simply is the derived category of $\mathbf{Z}_\ell$-modules of finite type, and we have

$$Tr(R_\ell(f)) = \sum_i (-1)^i Tr(H^i R_\ell(f) : H^i R_\ell(M) \to H^i R_\ell(M))$$

where each $Tr(H^i R_\ell(f))$ can be computed in the usual way in terms of traces of matrices. If k is not separably closed, we can always choose a separable closure $\bar{k}$ and observe that pulling back along the map $Spec(\bar{k}) \to Spec(k)$ is a symmetric monoidal functor which commutes with the ℓ-adic realization functor. This can actually be used to prove that the Euler characteristic is always an integer (as opposed to a rational number in $\mathbf{Z}[1/p]$): if $f = 1_M$ is the identity, the trace of $R_\ell(f)$ can be computed as an alternating sum of ranks of $\mathbf{Z}_\ell$-modules of finite type.

Corollary 3.3.2.7 *For any dualizable object M in $DM_h(k, \mathbf{Z})$, we have $\chi(M) \in \mathbf{Z}$.*

3.3.2.8 Let A be a ring. A function $f : X \to A$ from a topological space to a ring is constructible if there is a finite stratification of X by locally closed X_i such that each $f|_{X_i}$ is constant. We denote by $C(X, A)$ the ring of constructible functions with values in A on X. For a scheme X, we define $C(X, A) = C(|X|, A)$, where $|X|$ denotes the topological space underlying X.

3.3.2.9 Recall that, for a stable ∞-category C, we have its Grothendieck group $K_0(C)$: the free monoid generated by isomorphism classes $[x]$ of objects x of C, modulo the relations $[x] = [x'] + [x'']$ for each cofiber sequence $x' \to x \to x''$. In particular, we have the relations $0 = [0]$ and $[x] + [y] = [x \oplus y]$. This monoid turns out to be an abelian group with $-[x] = [\Sigma(x)]$. If ever C is symmetric monoidal, then $K_0(C)$ inherits a commutative ring structure with multiplication $[x][y] = [x \otimes y]$.

3.3.2.10 We have the Euler characteristic map $DM_{h,lc}(X, \mathbf{Z}) \xrightarrow{\chi} C(X, \mathbf{Z})$. It is defined by $\chi(M)(x) = \chi(x^*M)$, where the point x is seen as a map $x : Spec(\kappa(x)) \to X$. Recall that if $M \in DM_h(X, \Lambda)$ is locally constructible then there is $U \subseteq X$ open and dense such that $M|_{(X-U)_{red}}$ is locally constructible and $M|_U$ is dualizable. Therefore, by noetherian induction, we see that $\chi(M) : |X| \to \mathbf{Z}$ is a constructible function indeed. For any cofiber sequence of dualizable objects

$$M' \to M \to M'',$$

we have

$$\chi(M) = \chi(M') + \chi(M'').$$

Since $\chi(M \otimes N) = \chi(M)\chi(N)$, we have a morphism of rings:

$$\chi : K_0(DM_{h,lc}(X, \mathbf{Z})) \to C(X, \mathbf{Z}),$$

and we have a commutative triangle:

$$
\begin{array}{ccc}
K_0(DM_{h,lc}(X, \mathbf{Z})) & \xrightarrow{R_\ell} & K_0(D_c^b(X_{et}, \mathbf{Z}_\ell)) \\
& \searrow^{\chi} \quad \swarrow^{\chi} & \\
& C(X, \mathbf{Z}) &
\end{array}
$$

3.3.2.11 Given a stable ∞-category C, there is the full subcategory C_{tors} which consists of objects x such that there exists an integer n such that $n.1_x \cong 0$. One checks that C_{tors} is a thick subcategory of C and one defines the Verdier quotient $C \otimes \mathbf{Q} = C/C_{tors}$. All this is a fancy way to say that one defines $C \otimes \mathbf{Q}$ as the ∞-category with the same set of objects as C, such that $\pi_0 Map_C(x, y) \otimes \mathbf{Q} = \pi_0 Map_{C \otimes \mathbf{Q}}(x, y)$ for all x and y. This is how one defines ℓ-adic sheaves:

$$D_c^b(X_{et}, \mathbf{Q}_\ell) = D_c^b(X_{et}, \mathbf{Z}_\ell) \otimes \mathbf{Q}.$$

When it comes to motives, we can prove that, when X is noetherian of finite dimension, the canonical functor

$$DM_{h,lc}(X, \mathbf{Z}) \otimes \mathbf{Q} \to DM_{h,lc}(X, \mathbf{Q})$$

is fully faithful and almost an equivalence: a Morita equivalence. Since $DM_{h,lc}(X, \mathbf{Q})$ is idempotent complete, that means that any $\mathbf{Q}$-linear locally constructible motive is a direct factor of a $\mathbf{Z}$-linear one. Furthermore, one checks that $D_c^b(X_{et}, \mathbf{Q}_\ell)$ is idempotent complete (because it has a bounded t-structure), so that we get a $\mathbf{Q}$-linear ℓ-adic realization functor:

$$R_\ell : DM_{h,lc}(X, \mathbf{Q}) \to D_c^b(X_{et}, \mathbf{Q}_\ell)$$

which is completely determined by the fact that the following square commutes.

$$
\begin{array}{ccc}
DM_{h,lc}(X, \mathbf{Z}) & \xrightarrow{R_\ell} & D_c^b(X_{et}, \mathbf{Z}_\ell) \\
\downarrow & & \downarrow \\
DM_{h,lc}(X, \mathbf{Q}) & \xrightarrow{R_\ell} & D_c^b(X_{et}, \mathbf{Q}_\ell)
\end{array}
$$

The $\mathbf{Q}$-linear ℓ-adic realization functor commutes with the six operations if we restrict ourselves to quasi-excellent schemes over $\mathbf{Z}[1/\ell]$; see [13, 7.2.24].

We may see these realization functors as a categorified version of cycle class maps. Indeed, in view of the representability results such as Theorem 3.2.4.3, they induce the classical cycle class maps in ℓ-adic cohomomology: for a field k and a separated morphism of finite type $a : X \to \mathrm{Spec}(k)$, we have

$$H^0(\mathrm{Hom}_{DM_h(X, \mathbf{Q})}(\mathbf{Q}(n)[2n], a^!\mathbf{Q})) \xrightarrow{R_\ell} H^0(\mathrm{Hom}_{D_c^b(X_{et}, \mathbf{Q}_\ell)}(\mathbf{Q}_\ell(n)[2n], a^!\mathbf{Q}_\ell))$$

$$\| \wr \qquad\qquad\qquad\qquad\qquad\qquad\qquad\qquad\qquad\qquad \| \wr$$

$$CH_n(X) \otimes \mathbf{Q} \dashrightarrow H_c^{2n}(X_{et}, \mathbf{Q}_\ell(n))^\wedge$$

If X is regular (e.g. smooth) this gives by Poincaré duality the cycle class map:

$$CH^n(X) \to H^{2n}(X_{et}, \mathbf{Q}_\ell(n)).$$

One can lift these cycle class maps to integral coefficients using similar arguments from *cdh*-motives; see [12].

Theorem 3.3.2.12 *There is a canonical exact sequence of the form:*

$$K_0(DM_{h,lc}(X, \mathbf{Z})_{tors}) \to K_0(DM_{h,lc}(X, \mathbf{Z})) \to K_0(DM_{h,lc}(X, \mathbf{Q})) \to 0.$$

Proof Let $DM_h(X, \mathbf{Z})'$ be the smallest localizing subcategory of $DM_h(X, \mathbf{Z})$ generated by $DM_{h,lc}(X, \mathbf{Z})_{tors}$. We also define $D(X_{et}, \mathbf{Z})'$ as the smallest localizing subcateory of $D(X_{et}, \mathbf{Z})$ generated by objects of the form $j_!(\mathcal{F})$, where $j : U \to X$ is a dense open immersion and $\mathcal{F}$ is bounded with constructible cohomology sheaf, such that there is a prime p with the following two properties:

- $p.1_{\mathcal{F}} = 0$;
- p is invertible in $\mathcal{O}_U$.

Then a variant of the rigidity Theorem 3.2.4.4 (together with Remark 3.2.4.5) gives an equivalence of ∞-categories:

$$DM_h(X, \mathbf{Z})' \cong D(X_{et}, \mathbf{Z})'.$$

One then checks that the t-structure on $D(X_{et}, \mathbf{Z})'$ induces a bounded t-structure on $DM_{h,lc}(X, \mathbf{Z})_{tors}$ (with noetherian heart, since we get a Serre subcategory of constructible étale sheaves of abelian groups on X_{et}). Using the basic properties of non-connective K-theory [8, 15, 42], we see that we have an exact sequence

$$K_0(DM_{lc}(X)_{tors}) \to K_0(DM_{lc}(X)) \to K_0(DM_{lc}(X)_{\mathbf{Q}}) \to K_{-1}(DM_{lc}(X)_{tors}),$$

where $DM_{lc}(X) = DM_{h,lc}(X, \mathbf{Z})$ and $DM_{lc}(X)_{\mathbf{Q}} = DM_{h,lc}(X, \mathbf{Q})$. By virtue of a theorem of Antieau, Gepner, and Heller [2], the existence of a bounded t-structure with noetherian heart implies that $K_{-i}(DM_{h,lc}(X, \mathbf{Z})_{tors}) = 0$ for all $i > 0$. $\square$

Here is a rather concrete consequence (since $\chi(M) = 0$ for M in $DM_{h,lc}(X, \mathbf{Z})_{tors}$).

Corollary 3.3.2.13 *For any M in $DM_{h,lc}(X, \mathbf{Q})$, there exists M_0 in $DM_{h,lc}(X, \mathbf{Z})$, such that, for any point x in X, we have $\chi(x^*M) = \chi(x^*M_0)$.*

Remark 3.3.2.14 It is conjectured that there is a (nice) bounded t-structure on $DM_{h,lc}(X, \mathbf{Q})$. Since $DM_{h,lc}(X, \mathbf{Z})_{tors})$ has a bounded t-structure, this would imply the existence of a bounded t-structure on $DM_{h,lc}(X, \mathbf{Z})$, which, in turns would imply the vanishing of $K_{-1}(DM_{h,lc}(X, \mathbf{Z}))$ (see [2]). Such a vanishing would mean that all Verdier quotients of $DM_{h,lc}(X, \mathbf{Z})$ would be idempotent-complete (see [42, Remark 1 p. 103]). In particular, we would have an equivalence of ∞-categories $DM_{h,lc}(X, \mathbf{Z}) \otimes \mathbf{Q} = DM_{h,lc}(X, \mathbf{Q})$. The previous proposition is a virtual approximation of this expected equivalence.

3.3.2.15 Let R be a ring and let $W(R) = 1 + R[[t]]$ the set of power series with coefficients in R and leading term equal to 1. It has an abelian group structure defined by the multiplication of power series. And it has a unique multiplication $*$ such that $(1 + at) * (1 + bt) = 1 + abt$, turning $W(R)$ into a commutative ring: the ring of Witt vectors. We also have the subset $W(R)_{rat} \subseteq W(R)$ of rational functions, which one can prove to be a subring. Given a (stable) ∞-category C, we define

$$C^{\mathbf{N}} = \{\text{objects of } C \text{ equipped with an endomorphism}\}.$$

This is again a stable ∞-category. For $C = Perf(R)$ the ∞-category of perfect complexes on the ring R, we have an exact sequence

$$0 \to K_0(Perf(R)) \to K_0(Perf(R)^{\mathbf{N}}) \to W(R)_{rat} \to 0$$

where the first map sends a perfect complex of R-modules M to the class of M equipped with the zero map $0 : M \to M$, while the second maps sends $f : M \to M$

to $\det(1 - tf)$ (it is sufficient to check that these maps are well defined when M is a projective module of finite type, since these generate the K-groups); see [1]. The first map identifies $K_0(R)$ with an ideal of $K_0(Perf(R)^{\mathbf{N}})$ so that we really get an isomorphism of commutative rings:

$$K_0(Perf(R)^{\mathbf{N}})/K_0(Perf(R)) \cong W(R)_{rat} \,.$$

3.3.2.16 Let k be a field with a given algebraic closure $\bar{k}$, as well as prime number ℓ which is distinct from the characteristic of k. We observe that $D_c^b(\bar{k}, \mathbf{Q}_\ell)$ simply is the bounded derived category of complexes of finite dimensional $\mathbf{Q}_\ell$-vector spaces. We thus have a symmetric monoidal realization functor

$$DM_{h,lc}(k, \mathbf{Q}) \to D_c^b(k, \mathbf{Q}_\ell) \to D_c^b(\bar{k}, \mathbf{Q}_\ell) \cong Perf(\mathbf{Q}_\ell) \,.$$

This induces a functor

$$DM_{h,lc}(k, \mathbf{Q})^{\mathbf{N}} \to Perf(\mathbf{Q}_\ell)^{\mathbf{N}},$$

and thus a map

$$K_0(DM_{h,lc}(k, \mathbf{Q})^{\mathbf{N}}) \to K_0(Perf(\mathbf{Q}_\ell)^{\mathbf{N}})$$

inducing a ring homomorphism, the ℓ-adic Zeta function

$$Z_\ell : K_0(DM_{h,lc}(k, \mathbf{Q})^{\mathbf{N}})/K_0(DM_{h,lc}(k, \mathbf{Q})) \to W(\mathbf{Q}_\ell)_{rat} \subseteq 1 + \mathbf{Q}_\ell[[t]] \,.$$

On the other hand, for an endomorphism $f : M \to M$ in $DM_{h,lc}(X, \mathbf{Q})$, one defines its motivic Zeta function as follows

$$Z(M, f) = \exp\left(\sum_{n \geq 1} Tr(f^n) \frac{t^n}{n} \right) \in 1 + \mathbf{Q}[[t]] \,.$$

Basic linear algebra show that $Z(M, f) = Z_\ell(M, f)$ (see [1]). In particular, we see that the ℓ-adic Zeta function $Z_\ell(M, f)$ has rational coefficients and is independent of ℓ, while the motivic Zeta function $Z(M, f)$ is rational. In other words, we get a morphism of rings

$$Z : K_0(DM_{h,lc}(X, \mathbf{Q})^{\mathbf{N}})/K_0(DM_{h,lc}(X, \mathbf{Q})) \to W(\mathbf{Q})_{rat} \subset W(\mathbf{Q}) \,.$$

Concretely, if there is a cofiber sequence of motivic sheaves equipped with endomorphisms in the stable ∞-category $DM_{h,lc}(X, \mathbf{Q})^{\mathbf{N}}$

$$(M', f') \to (M, f) \to (M'', f'') \,,$$

then

$$Z(M, f)(t) = Z(M', f')(t) \cdot Z(M'', f'')(t)$$

holds in $\mathbf{Q}[[t]]$. And for two motivic sheaves equipped with endomorphisms (M, f) and (M', f') in $DM_{h,lc}(X, \mathbf{Q})^{\mathbf{N}}$, there is

$$Z(M \otimes M', f \otimes f') = Z(M, f) * Z(M', f')$$

where $*$ denotes the multiplication in the big ring of Witt vectors $W(\mathbf{Q})$.

3.3.2.17 Take $k = \mathbf{F}_q$ a finite field and let $M_0 \in DM_{h,lc}(k, \mathbf{Q})$, with $M = p^* M_0$, $p : spec(\bar{k}) \to spec(k)$. Let $F : M \to M$ be the induced Frobenius. We define the *Riemann-Weil Zeta function* of M_0 as:

$$\zeta(M_0, s) = Z(M, F)(t), \quad t = p^{-s}.$$

The fact that the assignment $Z(-, F)$ defines a morphism of rings with values in $W(\mathbf{Q})$ can be used to compute explicitly the Zeta function of many basic schemes such as $\mathbf{P}^n$ of $(\mathbf{G}_m)^r$; see [37, Remark 2.2] for instance.

3.3.3   Grothendieck-Verdier Duality

3.3.3.1 Take S be a quasi-excellent regular scheme. We choose a $\otimes$-invertible object I_S in $DM_h(S, \Lambda)$ (e.g. $I_S = \mathbf{Z}(d)[2d]$, where d is the Krull dimension of S). For $a : X \to S$ separated of finite type, we define $I_X = a^! I_S$.

Define $\mathbf{D}_X : DM_h(X, \Lambda)^{op} \to DM_h(X, \Lambda)$ by

$$\mathbf{D}_X(M) = Hom(M, I_X).$$

We will sometimes write $\mathbf{D}(M) = \mathbf{D}_X(M)$.

Theorem 3.3.3.2 *For M a locally constructible motivic sheaf over X, the canonical map $M \to \mathbf{D}_X \mathbf{D}_X(M)$ is an equivalence.*

There is a proof in the literature under the additional assumption that S is of finite type over an excellent scheme of dimension ≤ 2 (see [13, 14]). But there is in fact a proof which avoids this extra hypothesis using higher categories. Here is a sketch.

Proof The formation of the Verdier dual is compatible with pulling back along an étale map. We may thus assume that M is constructible. The full subcategory of those M's such that the biduality map of the theorem is invertible is thick. Therefore, we may assume that $M = M(U)$ for some smooth X-scheme U. In particular, we may assume that $M = \Lambda \otimes \Sigma^\infty \mathbf{Z}(U)$. It is thus sufficient to prove the case where $\Lambda = \mathbf{Z}$. By standard arguments, we see that is sufficient to prove the case where Λ

is finite or $\Lambda = \mathbf{Q}$. Such duality theorem is a result of Gabber [26] for the derived category of sheaves on the small étale site of X with coefficients in Λ of positive characteristic with n invertible in $\mathcal{O}_X$. By Theorem 3.2.4.4 and Remark 3.2.4.5, this settles the case where Λ is finite. It remains to prove the case where $\Lambda = \mathbf{Q}$. We will first prove the following statement. For each separated morphism of finite type $a : X \to S$, and each integer n, the natural map

$$\mathrm{Hom}_{DM_h(X,\mathbf{Q})}(\mathbf{Q}, \mathbf{Q}(n)) \to \mathrm{Hom}_{DM_h(X,\mathbf{Q})}(I_X, I_X(n))$$

is invertible in $D(\mathbf{Q})$ (this is the map obtained by applying the global section functor $\mathrm{Hom}(\mathbf{Q}, -)$ to the unit map $\mathbf{Q} \to Hom(I_X, I_X)$). We observe that we may see this map as a morphism of presheaves of complexes of $\mathbf{Q}$-vector spaces

$$E \to F$$

where $E(X) = \mathrm{Hom}_{DM_h(X,\mathbf{Q})}(\mathbf{Q}, \mathbf{Q}(n))$ and $F(X) = \mathrm{Hom}_{DM_h(X,\mathbf{Q})}(I_X, I_X(n))$. For a morphism of S-schemes $f : X \to Y$, the induced map $E(Y) \to E(X)$ is induced by the functor f^*, while the induced map $F(Y) \to F(X)$ is induced by the functor $f^!$ (and the fact that $f^!(I_Y) \cong I_X$).[5] Now, we observe that both E and F are in fact h-sheaves of complexes of $\mathbf{Q}$-vector spaces. Indeed, using [14, Proposition 3.3.4], we see that E and F satisfy Nisnevich excision and thus are Nisnevich sheaves. On the other hand, one can also characterise h-descent for $\mathbf{Q}$-linear Nisnevich sheaves by suitable excision properties [14, Theorem 3.3.24]. Such properties for E and F follow right away from [14, Theorem 14.3.7 and Remark 14.3.38], which proves the property of h-descent for E and F. By virtue of Theorem 3.3.1.10, it is sufficient to prove that $E(X) \cong F(X)$ for X regular and affine. In particular, $a : X \to S$ factors through a closed immersion $i : X \to \mathbf{A}^n \times S$. By relative purity, we have

$$I_{\mathbf{A}^n \times S} \cong p^*(I_S)(n)[2n]$$

and thus $I_{\mathbf{A}^n \times S}$ is $\otimes$-invertible (where $p : \mathbf{A}^n \times S \to S$ is the second projection). This implies that

$$I_X \cong i^!(I_{\mathbf{A}^n \times S}) \cong i^!(\mathbf{Q}) \otimes i^*(I_{\mathbf{A}^n \times S})$$

(Hint: use the fact that $i^! Hom(A, B) \cong Hom(i^* A, i^! B)$). By Absolute Purity, we have $i^! \mathbf{Q} \cong \mathbf{Q}(-c)[-2c]$, where c is the codimenion of i. In particular, the object I_X is $\otimes$-invertible, and thus the unit map $\mathbf{Q} \to Hom(I_X, I_X)$ is invertible. This implies that the map $E(X) \to F(X)$ is invertible as well.

---

[5]This is where ∞-category theory appears seriously: proving that the construction $f \mapsto f^!$ actually defines a presheaf is a highly non-trivial homotopy coherence problem. Such construction is explained in [38, Chapter 10], using the general results of [29, 30].

We will now prove that the unit map

$$\mathbf{Q} \to Hom(I_X, I_X)$$

is invertible in $DM_h(X, \mathbf{Q})$ for any separated S-scheme of finite type X. Equivalently, we have to prove that, for any smooth X-scheme U and any integer n, the induced map

$$\mathrm{Hom}_{DM_h(X,\mathbf{Q})}(M(U), \mathbf{Q}(n)) \to \mathrm{Hom}_{DM_h(X,\mathbf{Q})}(M(U), Hom(I_X, I_X)(n))$$

is invertible in $D(\mathbf{Q})$. But we have

$$\mathrm{Hom}(f_\sharp \mathbf{Q}, \mathbf{Q}(n)) \cong \mathrm{Hom}(\mathbf{Q}, \mathbf{Q}(n)) = E(U)$$

with a smooth structural map $f : U \to X$, and

$$\mathrm{Hom}(f_\sharp \mathbf{Q}, Hom(I_X, I_X)(n)) \cong \mathrm{Hom}(\mathbf{Q}, f^* Hom(I_X, I_X)(n))$$

$$\cong \mathrm{Hom}(\mathbf{Q}, Hom(f^* I_X, f^* I_X)(n))$$

$$\cong \mathrm{Hom}(\mathbf{Q}, Hom(f^! I_X(-d)[-2d], f^! I_X(-d)[-2d])(n))$$

$$\cong \mathrm{Hom}(\mathbf{Q}, Hom(f^! I_X, f^! I_X)(n))$$

$$\cong \mathrm{Hom}(\mathbf{Q}, Hom(I_U, I_U)(n)) = F(U).$$

In other words, we just have to check that the map $E(U) \to F(U)$ is invertible, which we already know.

Finally, we can prove that the canonical map $M \to \mathbf{D}_X \mathbf{D}_X(M)$ is invertible. As already explained at the beginning of the proof, it is sufficient to prove this when M is constructible. By virtue of Proposition 3.3.1.7, it is sufficient to prove the case where $M = f_*(\mathbf{Q})$, for $f : Y \to X$ a proper map. We have:

$$\mathbf{D}_X f_* \mathbf{Q} = Hom(f_* \mathbf{Q}, I_X)$$

$$\cong f_* Hom(\mathbf{Q}, f^! I_X)$$

$$\cong f_* f^! I_X$$

$$\cong f_* I_Y.$$

Therefore, we have

$$\mathbf{D}_X \mathbf{D}_X(M) \cong \mathbf{D}_X f_* I_Y$$

$$\cong Hom(f_* I_Y, I_X)$$

$$\cong f_* Hom(I_Y, f^! I_X)$$

$$\cong f_* Hom(I_Y, I_Y)$$

$$\cong f_* \mathbf{Q} = M,$$

and this ends the proof. □

Corollary 3.3.3.3 *For locally constructible motives and f a morphism between separated S-schemes of finite type, we have:*

$$\mathbf{D} f_* \cong f_! \mathbf{D}$$

$$\mathbf{D} f_! \cong f_* \mathbf{D}$$

$$\mathbf{D} f^! \cong f^* \mathbf{D}$$

$$\mathbf{D} f^* \cong f^! \mathbf{D}.$$

(The proof is by showing tautologically the second one and the fourth one, and then deduce the other two using that $\mathbf{D}$ is an involution.)

Proposition 3.3.3.4 *For any M and N in $DM_h(X, \Lambda)$, if N is locally constructible, then*

$$\mathbf{D}(M \otimes \mathbf{D}N) \cong Hom(M, N).$$

Proof We construct a canonical comparison morphism:

$$Hom(M, N) \to \mathbf{D}(M \otimes \mathbf{D}N).$$

By transposition, it corresponds to a map

$$M \otimes Hom(M, N) \otimes \mathbf{D}(N) \to I_X.$$

Such a map is induced by the evaluation maps

$$M \otimes Hom(M, N) \to N \quad \text{and} \quad N \otimes \mathbf{D}(N) \to I_X.$$

For N fixed, the class of M's such that this map is invertible is closed under colimits. Therefore, we reduce the question to the case where $M = f_\sharp \Lambda$ for $f : X \to S$ a smooth map of dimension d. In that case, we have

$$Hom(M, N) \cong f_* f^*(N),$$

while

$$\mathbf{D}(M \otimes \mathbf{D}N) \cong \mathbf{D}(f_! f^*(\Lambda(-d)[-2d]) \otimes \mathbf{D}N)$$

$$\cong \mathbf{D}(f_! f^*(\Lambda) \otimes \mathbf{D}N)(d)[2d]$$

$$\cong \mathbf{D}(f_! f^*(\mathbf{D}N))(d)[2d]$$

$$\cong f_* f^!(\mathbf{D}\mathbf{D}N)(d)[2d]$$

$$\cong f_* f^* N,$$

which ends the proof. $\square$

Corollary 3.3.3.5 *For M and N locally constructible on X, we have:*

$$M \otimes N \cong \mathbf{D}Hom(M, \mathbf{D}N).$$

3.3.4 Generic Base Change: A Motivic Variation on Deligne's Proof

3.3.4.1 The following statement, is a motivic analogue of Deligne's generic base change theorem for torsion étale sheaves [17, Th. Finitude, 1.9]. The proof follows essentially the same pattern as Deligne's original argument, except that locally constant sheaves are replaced by dualizable objects, as we will explain below. We will write $DM_h(X) = DM_h(X, \Lambda)$ for some fixed choice of coefficient ring Λ.

Theorem 3.3.4.2 (Motivic Generic Base Change Formula) *Let $f : X \to Y$ be a morphism between separated schemes of finite type over a noetherian base scheme S. Let M be a locally constructible h-motive on X. Then there is a dense subscheme $U \subset S$ such that the formation of $f_*(M)$ is compatible with any base-change which factors through U. Namely, for each $w : S' \to S$ factoring through U we have*

$$v^* f_* M \cong f'_* u^* M$$

where

$$
\begin{array}{ccc}
X' & \xrightarrow{u} & X \\
\downarrow{\scriptstyle f'} & & \downarrow{\scriptstyle f} \\
Y' & \xrightarrow{v} & Y \\
\downarrow & & \downarrow \\
S' & \xrightarrow{w} & S
\end{array}
$$

is the associated pull-back diagram.

Remark 3.3.4.3 The motivic generic base change formula is also a kind of independence of ℓ-result: for each prime ℓ so that the ℓ-adic realization is defined, the formation of $f_* R_\ell(M) \cong R_\ell(f_* M)$ is compatible with any base change over $U \subset S$, where U is a dense open subscheme which is given independently of ℓ.

The first step in the proof of Theorem 3.3.4.2 is to find sufficient conditions for the formation a direct image to be compatible with arbitrary base change.

Proposition 3.3.4.4 *Let $f : X \to S$ be a smooth morphism of finite type between noetherian schemes, and let us consider a locally constructible h-motive M over X. Assume that M is dualizable in $DM_{h,lc}(X)$ and that the direct image with compact support of its dual $f_!(M^\wedge)$ is dualizable as well in $DM_{h,lc}(S)$. Then $f_*(M)$ is* *dualizable (in particular, locally constructible), and, for any pullback square of the form*

$$
\begin{array}{ccc}
X' & \xrightarrow{\ u\ } & X \\
{\scriptstyle f'}\downarrow & & \downarrow{\scriptstyle f} \\
S' & \xrightarrow{\ v\ } & S
\end{array}
$$

the morphism f' is smooth, the pullback $u^*(M)$ is dualizable, so is $f'_!(u^*(M)^\wedge)$, and, furthermore, the canonical base change map $v^* f_*(M) \rightarrow f'_* u^*(M)$ is invertible.

Proof If d denotes the relative dimension of X over S (seen as a locally constant function over S), we have:

$$
f_*(M) \simeq f_* Hom(M^\wedge, \Lambda)
$$

$$
\simeq f_* Hom(M^\wedge, f^! \Lambda)(-d)[-2d]
$$

$$
\simeq Hom(f_!(M^\wedge), \Lambda)(-d)[-2d]
$$

$$
\simeq (f_!(M^\wedge))^\wedge(-d)[-2d]
$$

(where the dual of a dualizable object A is denoted by $A^\wedge$). Remark that pullback functors v^* are symmetric monoidal and thus preserve dualizable objects as well as the formation of their duals. Therefore, for any pullback square of the form

$$
\begin{array}{ccc}
X' & \xrightarrow{\ u\ } & X \\
{\scriptstyle f'}\downarrow & & \downarrow{\scriptstyle f} \\
S' & \xrightarrow{\ v\ } & S
\end{array}
$$

we have that f' is smooth of relative dimension d, that $u^*(M)$ is dualizable with dual $u^*(M)^\wedge \simeq u^*(M^\wedge)$, and:

$$
v^* f_*(M) \simeq v^*(f_!(M^\wedge))^\wedge(-d)[-2d]
$$

$$
\simeq (v^* f_!(M^\wedge))^\wedge(-d)[-2d]
$$

$$
\simeq (f'_! u^*(M^\wedge))^\wedge(-d)[-2d]
$$

$$
\simeq (f'_!(u^*(M)^\wedge))^\wedge(-d)[-2d]
$$

This also shows that $f'_!(u^*(M)^\wedge)$ is dualizable and thus that there is a canonical isomorphism

$$
(f'_!(u^*(M)^\wedge))^\wedge(-d)[-2d] \simeq f'_*(u^*(M)).
$$

We deduce right away from there that the canonical base change map $v^* f_*(M) \to f'_*(u^*(M))$ is invertible.   □

Remark 3.3.4.5 In the preceding proposition, we did not use any particular property of $DM_{h,lc}$: the statement and its proof hold in any context in which we have the six operations (more precisely, we mainly used the relative purity theorem as well as the proper base change theorem).

In order to prove Theorem 3.3.4.2 in general, we need to verify the following property of h-motives.

Proposition 3.3.4.6 *Let S be a noetherian scheme of finite dimension, and $f : Y \to S$ a quasi-finite morphism of finite type. The functors $f_! : DM_h(X) \to DM_h(S)$ and $f_* : DM_h(X) \to DM_h(S)$ are conservative.*

Proof If f is an immersion, then $f_!$ and f_* are fully faithful, hence conservative. Since the composition of two conservative functors is conservative, Zariski's Main Theorem implies that it is sufficient to prove the case where f is finite. In this case, since the formation of $f_! \simeq f_*$ commutes with base change along any map $S' \to S$, by noetherian induction, it is sufficient to prove this assertion after restricting to a dense open subscheme of S of our choice. Since, for h-motives, pulling back along a surjective étale morphism is conservative, we may even replace S by an étale neighbourhood of its generic points. For f surjective and radicial, [13, Proposition 6.3.16] ensures that $f_!$ is an equivalence of categories. We may thus assume that f also is étale. If ever $X = X' \amalg X''$, and if f' and f'' are the restriction of f to X' and X'', respectively, then we have $DM_h(X) \simeq DM_h(X') \times DM_h(X'')$, and the functor $f_!$ decomposes into

$$f_!(M) = f'_!(M') \oplus f''_!(M'')$$

for $M = (M', M'')$. Therefore, it is then sufficient to prove the proposition for f' and f'' separately. Replacing S by an étale neighbourhood of its generic points, we may thus assume that either X is empty, either f is an isomorphism, in which cases the assertion is trivial.   □

3.3.4.7 Let $P(n)$ be the assertion that, whenever S is integral and $f : X \to Y$ is a separated morphism of S-schemes of finite type, such that the dimension of the generic fiber of X over S is smaller than or equal to n, then, for any locally constructible h-motive M on X, there is a dense open subscheme U of S such that the formation of $f_*(M)$ is compatible with base change along maps $S' \to U \subset S$.

From now on, we fix a separated morphism of S-schemes of finite type $f : X \to Y$; as well as a locally constructible h-motive M on X.

Lemma 3.3.4.8 *The property that there exists a dense open subscheme $U \subset S$ such that the formation of $f_*(M)$ is stable under any base change along maps $S' \to U \subset S$ is local on Y for the Zariski topology.*

Proof Indeed, assume that there is an open covering $Y = \bigcup_j V_i$ such that, for each j, there is a dense open subset $U_j \subset U$ with the property that the formation of the motive $(f^{-1}(V_j) \to V_j)_*(M_{f^{-1}(V_j)})$ is stable under any base change along maps of the form $S' \to U_j \subset S$. Since Y is noetherian, we may assume that there finitely many V_j's, so that $U = \bigcap_j U_j$ is a dense open subscheme of S. For any j, the formation of $(f^{-1}(V_j) \to V_j)_*(M_{f^{-1}(V_j)})$ is stable under any base change along maps of the form $S' \to U \subset S$. Since pulling back along open immersions commutes with any push-forward, one deduces easily that the formation of $f_*(M)$ is stable under any base change of the form $(f^{-1}(V_j) \to V_j)_*(M_{f^{-1}(V_j)})$ is stable under any base change along maps of the form $S' \to U \subset S$.   □

Lemma 3.3.4.9 *Assume that there is a compactification of Y: an open immersion $j : Y \to \bar{Y}$ with $\bar{Y}$ a proper S-scheme. If there is a dense open subscheme U such that the formation of $(jf)_*(M)$ is compatible with all base changes along maps $S' \to U \subset S$, then the formation of $f_*(M)$ is compatible with all base changes along maps $S' \to U \subset S$.*

Proof This follows right away from the fact that pulling back along j is compatible with any base changes and from the fully faithfulness of the functor j_* (so that $j^* j_* f_*(M) \simeq f_*(M)$).   □

Lemma 3.3.4.10 *Assume that S is integral, that the dimension of the generic fiber of X over S is $n \geq 0$, and that $P(n-1)$ holds. If X is smooth over S, and if M is dualizable, then there is a dense open subscheme of S such that the formation of $f_*(M)$ is stable under base change along maps $S' \to U \subset S$.*

Proof Since pulling back along open immersions commutes with any push-forward, and since Y is quasi-compact, the problem is local over Y. Therefore, we may assume that Y is affine. Let us choose a closed embedding $Y \subset \mathbf{A}_S^d$ determined by d functions $g_i : Y \to \mathbf{A}_S^1$, $1 \leq i \leq d$. For each index i, we may apply $P(n-1)$ to f, seen as open embedding of schemes over $\mathbf{A}_S^1$ through the structural map g_i. This provides a dense open subscheme U_i in $\mathbf{A}_S^1$ such that the formation of $f_*(M)$ is compatible with any base change of g_i along a map $S' \to \mathbf{A}_S^1$ which factors through U_i. Let V be the union of all the open subschemes $g_i^{-1}(U_i)$, $1 \leq i \leq d$, and let us write $j : V \to Y$ for the corresponding open immersion. Then the formation of $j_! j^* f_*(M)$ is compatible with any base change $S' \to S$. Let us choose a closed complement $i : T \to Y$ to j. Then T is finite: the reduced geometric fibers of T/S are traces on Y of the subvarieties of $\mathbf{A}^d$ determined by the vanishing of all the non constant polynomials $p_i(x_i) = 0$, $1 \leq i \leq d$, where $p_i(x)$ is a polynomial such that $U_i = \{p_i(x) \neq 0\}$.

We may now consider the closure $\bar{Y}$ of Y in $\mathbf{P}_S^n$. Any complement of V in $\bar{Y}$ is also finite over a dense open subscheme of S: the image in S of the complement of V in $\bar{V}$ is closed (since $\bar{V}$ is proper over S), and does not contain the generic point (since the generic fiber of X is not empty), so that we may replace S by the complement of this image. By virtue of Lemma 3.3.4.9, we may replace Y by $\bar{Y}$, so that we are reduced to the following situation: the scheme Y is proper over S, and there is a dense open immersion $j : V \to Y$ with the property that the formation of $j_! j^* f_*(M)$ is compatible with any base change $S' \to S$, and that after shrinking S, there is a closed complement $t : T \to Y$ of V which is finite over S. We thus have the following canonical cofiber sequence

$$j_! j^* f_*(M) \to f_*(M) \to i_* i^* f_*(M)$$

Let $p : Y \to S$ be the structural map (which is now proper). We already know that the formation of $j_! j^* f_*(M)$ is compatible with any base change of the form $S' \to S$. Therefore, it is sufficient to prove that, possibly after shrinking S, the formation of $i_* i^* f_*(M)$ has the same property. Since $i_! \simeq i_*$, this means that this is equivalent to the property that, possibly after shrinking S, the formation of $i^* f_*(M)$ is compatible with any base change of the form $S' \to S$. But the composed morphism pi being finite, by virtue of Proposition 3.3.4.6, we are reduced to prove this property for $p_* i_* i^* f_*(M)$. We then have the following canonical cofiber sequence

$$p_* j_! j^* f_*(M) \to (pf)_*(M) \to (pi)_* i^* f_*(M)$$

By virtue of Proposition 3.3.4.4, possibly after shrinking S, the formation of $(pf)_*(M)$ is compatible with any base change. Since p is proper, we have the proper base change formula (because $p_! \simeq p_*$), and therefore, the formation of $j_! j^* f_*(M)$ being compatible with any base change of the form $S' \to S$, the formation of $p_* j_! j^* f_*(M)$ is also compatible with any base change $S' \to S$. One deduces that, possibly after shrinking S the fomration of $(pi)_* i^* f_*(M)$ is also compatible with any base change $S' \to S$. $\qquad\square$

Proof of Theorem 3.3.4.2 We observe easily that it is sufficient to prove the case where S is integral. We shall prove $P(n)$ by induction. The case $n = -1$ is clear. We may thus assume that $n \geq 0$ and that $P(n - 1)$ holds true. Locally for the h-topology, radicial surjective and integral morphisms are isomorphisms; in particular, pulling back along a radicial surjective and integral morphism is an equivalence of categories which commutes with the six operations. There is a dense open subscheme U of S and a finite radicial and surjective map $U' \to U$, so that $X' = X \times_S U'$ has a dense open subscheme which is smooth over U' (it is sufficent to prove this over the spectrum of the field of functions of S, by standard limit arguments). Replacing S by U' and X by X', we may thus assume, without loss of generality, that the smooth locus of X over S is a dense open subscheme.

Let $j : V \to X$ be a dense open immersion such that V is smooth over S. Shrinking V, we may assume furthermore that $M_{|V}$ is dualizable in $DM_h(V)$. We choose a closed complement $i : Z \to X$ of V. With $N = i^!(M)$, we then have the following canonical cofiber sequence:

$$i_*(N) \to M \to j_* j^*(M)$$

By virtue of Lemma 3.3.4.10, possibly after shrinking S, we may assume that the formation of $j_*(M)$ is compatible with base changes along maps $S' \to S$. So is the formation of $i_*(N)$, since i is proper. Applying the functor f_* to the distinguished triangle above, we obtain the following cofiber sequence:

$$(fi)_*(N) \to f_*(M) \to (fj)_* j^*(M) .$$

We may apply Lemma 3.3.4.10 to fj and M, and observe that $P(n-1)$ applies to fi and N. Therefore, there exists a dense open subscheme $U \subset S$ such that the formation of $(fi)_*(N)$ and of $(fj)_* j^*(M)$ is compatible with any base change along maps $S' \to U \subset S$. This implies that the formation of $f_*(M)$ is compatible with such base changes as well.   □

3.4   Characteristic Classes

3.4.1   Künneth Formula

3.4.1.1 Let k be a field. All schemes will be assumed to be separated of finite type over k.

Theorem 3.4.1.2 *Let $f : X \to Y$ be a map of schemes, and T a scheme. Consider the square*

$$\begin{array}{ccc} T \times X & \xrightarrow{\ pr_2\ } & X \\ {\scriptstyle 1 \times f}\downarrow & & \downarrow{\scriptstyle f} \\ T \times Y & \xrightarrow{\ pr_2\ } & Y \end{array}$$

obtained by multiplying $f : X \to Y$ and $T \to Spec(k)$. Then $pr_2^* f_* \cong (1 \times f)_* pr_2^*$ holds.

Proof Since, for a field k and $S = Spec(k)$, the only dense open subscheme of S is S itself, the generic base change formula gives that the canonical map $pr_2^* f_*(M) \to (1 \times f)_* pr_2^*(M)$ is an isomorphism for any locally constructible motive M on X. Since we are comparing colimit preserving functors and since any motive is a colimit of locally constructible ones, this proves the theorem.   □

Some consequences:

1. Take X, T to be schemes and $pr_2 : T \times X \to X$ the projection. Then, for any M locally constructible on X we have:

$$pr_2^* Hom(M, N) \cong Hom(pr_2^* M, pr_2^* N) .$$

It is proved by producing a canonical map and then prove for a fixed N and reduce to the case where M is a generator, namely $M = f_\sharp \Lambda$ for smooth f. Then we get $Hom(M, N) \cong f_* f^* M$.

2. For a morphism $f : X \to Y$ consider the square below.

$$
\begin{array}{ccc}
T \times X & \xrightarrow{\;pr_2\;} & X \\
\downarrow{\scriptstyle 1 \times f} & & \downarrow{\scriptstyle f} \\
T \times Y & \xrightarrow{\;pr_2\;} & Y
\end{array}
$$

Then $pr_2^* f^! \cong (1 \times f)^! pr_2^*$.

For the proof observe that this is a local problem so that we may assume f is quasi-projective. The map f then has a factorization $f = g \circ i \circ j$ where g is smooth, i is a closed immersion, and j is an open immersion. Then $j^* = j^!$ and $g^* = g^!(-d)[-2d]$ so we reduce to the case where f is a closed immersion. Then f_* and $(1 \times f)_*$ are fully faithfull hence conservative. Therefore, it suffices to show

$$(1 \times f)_* pr_2^* f^! \cong (1 \times f)_* (1 \times f)^! pr_2^*.$$

Since left hand side is isomorphic to

$$pr_2^* f_* f^! ,$$

we only need to commute $f_* f^!$ and pr_2^*. Now observe that $f_* f^!(M) \cong Hom(f_* \Lambda, M)$. So we deduce the commutation of $f_* f^!$ from the commutation with internal Hom and f_* (which we both know). We finally have proper base change $pr_2^* f_*(\Lambda) \cong (1 \times f)_* pr_2^*$ and this finishes the proof.

Remark 3.4.1.3 If f is smooth or M is 'smooth' (dualizable) then for all N we have

$$f^* Hom(M, N) \cong Hom(f^* M, f^* N)$$

(see Remark 3.3.1.2).

3.4.1.4 For X a scheme and $a : X \to Spec(k)$ we define the dualizing sheaf to be $I_X = a^! \Lambda$ and $\mathbf{D}_X = Hom(-, I_X)$. If X, Y are schemes we can consider their product $X \times Y$ with projections $p_X : X \times Y \to X$ and $p_Y : X \times Y \to Y$. If M, N are motivic sheaves on X, Y respectively, we can define

$$M \boxtimes N := p_X^* M \otimes p_Y^* N$$

and then, recalling that $A \otimes B \cong \mathbf{D}Hom(A, \mathbf{D}B)$, we get that

$$M \boxtimes N \cong \mathbf{D}(Hom(p_X^* M, \mathbf{D}p_Y^* N)) \cong \mathbf{D}Hom(p_X^* M, p_Y^! \mathbf{D}N)$$

and therefore

$$M \boxtimes \mathbf{D}N \cong \mathbf{D}Hom(p_X^* M, p_Y^! N)$$

Theorem 3.4.1.5 *Let* X, Y *be schemes and* N *locally constructible on* Y. *Then* $p_Y^! N \cong I_X \boxtimes N$.

Proof Let a_X and a_Y be the structure maps of X, Y to $Spec(k)$. Then

$$p_X^* a_X^! \cong p_Y^! a_Y^*.$$

We have $I_X = a_X^!(\Lambda)$ and $p_X^*(I_X) \cong p_X^!(\Lambda)$. Moreover:

$$p_X^*(I_X) \cong p_X^*(\mathbf{D}_X \Lambda) \cong \mathbf{D}_{X \times Y} p_X^* \Lambda \cong \mathbf{D}_{X \times Y} p_Y^*(I_Y).$$

Then we have

$$
\begin{aligned}
p_X^* I_X \otimes p_Y^* N &\cong \mathbf{D} p_Y^* I_Y \otimes p_Y^* N \\
&\cong \mathbf{D}Hom(p_Y^* N, p_Y^* I_Y) \\
&\cong \mathbf{D} p_Y^* Hom(N, I_Y) \\
&\cong \mathbf{D} p_Y^* \mathbf{D}N \\
&\cong p_Y^! N
\end{aligned}
$$

Hence $I_X \boxtimes N \cong p_Y^! N$.   □

Corollary 3.4.1.6 $I_X \boxtimes I_Y \cong I_{X \times Y}$.

Proof $I_{X \times Y} \cong p_Y^! a_Y^! \Lambda \cong p_Y^! I_Y \cong I_X \boxtimes I_Y$.   □

Proposition 3.4.1.7 (Künneth Formula with Compact Support) *Let* $f : U \to X$ *and* $g : V \to Y$ *and let* $A \in DM_h(U, \Lambda)$ *and* $B \in DM_h(V, \Lambda)$ *then*

$$f_!(M) \boxtimes g_!(N) \cong (f \times g)_!(M \boxtimes N).$$

Proof Since $(f \times g)_! \cong (f \times 1)_!(1 \times g)_!$, we see that it is sufficient to prove this when f or g is the identity. Using the functorialities induced by permuting the factors $X \times Y \cong Y \times X$, we see that it is sufficient to prove the case where g is the identity. We then have a Cartesian square

$$
\begin{array}{ccc}
U \times Y & \xrightarrow{p_U} & U \\
{\scriptstyle f \times 1} \downarrow & & \downarrow {\scriptstyle f} \\
X \times Y & \xrightarrow{p_X} & X
\end{array}
$$

inducing an isomorphism

$$(f \times 1)_! p_U^* \cong p_X^* f_! .$$

The projection formula also gives

$$(f \times 1)_! (p_U^*(M)) \otimes p_Y^*(N) \cong (f \times 1)_! (M \boxtimes N)$$

so that we get $f_!(M) \boxtimes N \cong (f \times 1)_!(M \boxtimes N)$. $\qquad\qquad\square$

Corollary 3.4.1.8 *For $X = Y$ we get $f_!(M) \otimes g_!(N) \cong \pi_! i^*(M \boxtimes N)$ where π : $U \times_X V \to X$ is the canonical map, while $i : U \times_X V \to U \times V$ is the inclusion map.*

Remark 3.4.1.9 For f, g proper we get $f_* M \boxtimes f_* N \cong (f \times g)_*(M \boxtimes N)$.

Theorem 3.4.1.10 *For $M \in DM_{h,lc}(X, \Lambda)$ and $N \in DM_{h,lc}(Y, \Lambda)$ we have*

$$\mathbf{D}(M \boxtimes N) \cong \mathbf{D}M \boxtimes \mathbf{D}N .$$

Proof We may assume that $M = f_*\Lambda$ and $N = g_*\Lambda$ with f, g proper. Then

$$\begin{aligned}
\mathbf{D}f_*\Lambda \boxtimes \mathbf{D}g_*\Lambda &\cong f_* I_U \boxtimes g_* I_V \\
&\cong (f \times g)_*(I_U \boxtimes I_V) \\
&\cong (f \times g)_* I_{U \times V} \\
&\cong \mathbf{D}((f \times g)_*\Lambda) \\
&\cong \mathbf{D}(f_*\Lambda \boxtimes g_*\Lambda) \\
&\cong \mathbf{D}(M \boxtimes N).
\end{aligned}$$

Hence $\mathbf{D}(M \boxtimes N) \cong \mathbf{D}M \boxtimes \mathbf{D}N$. $\qquad\qquad\square$

Corollary 3.4.1.11 $\mathbf{D}M \boxtimes N \cong Hom(p_X^* M, p_Y^! N)$ *for M and N locally constructible.*

Corollary 3.4.1.12 (Künneth Formula in Cohomology) *Let us consider f : $U \to X$ and $g : V \to Y$ together with $M \in DM_h(U, \Lambda)$ and $N \in DM_h(V, \Lambda)$. Then*

$$f_*(M) \boxtimes g_*(N) \cong (f \times g)_*(M \boxtimes N).$$

Proof Functors of the form p_*, for p separated of finite type, commute with small colimits: since they are exact, it is sufficient to prove that they commute with small sums, which follows from [13, Prop. 5.5.10]. Therefore it is sufficient to prove this when M and N are (locally) constructible. In this case, the series of isomorphisms

$$f_*(M) \boxtimes g_*(N) \cong \mathbf{DD}(f_*(M) \boxtimes g_*(N))$$

$$\cong \mathbf{D}(\mathbf{D}f_*(M) \boxtimes \mathbf{D}g_*(N))$$

$$\cong \mathbf{D}(f_! \mathbf{D}M \boxtimes g_! \mathbf{D}N)$$

$$\cong \mathbf{D}((f \times g)_!(\mathbf{D}M \boxtimes \mathbf{D}N))$$

$$\cong \mathbf{D}((f \times g)_!(\mathbf{D}(M \boxtimes N))$$

$$\cong \mathbf{DD}((f \times g)_*(M \boxtimes N))$$

$$\cong (f \times g)_*(M \boxtimes N)$$

proves the claim.   □

Remark 3.4.1.13 In the situation of the previous corollary, if $X = Y = Spec(k)$, then also $X \times Y = Spec(k)$, so that the exterior tensor product $\boxtimes$ in $DM_h(X \times Y, \Lambda)$ simply corresponds to the usual tensor product $\otimes$ on $DM_h(k, \Lambda)$. We thus get a Künneth formula of the form

$$(a_U)_*(M) \otimes (a_V)_*(N) \cong (a_U \times a_V)_*(M \boxtimes N) .$$

Corollary 3.4.1.14 *Let us consider* $f : U \to X$ *and* $g : V \to Y$, *together with* $M \in DM_{h,lc}(X, \Lambda)$ *and* $N \in DM_{h,lc}(Y, \Lambda)$. *Then*

$$f^!(M) \boxtimes g^!(N) \cong (f \times g)^!(M \boxtimes N).$$

Proof For any separated morphism of finite a, the functor $a^!$ commutes with small colimits (since, they are exact, it is sufficient to prove that they commutes with small sums, which is asserted by Cisinski and Déglise [13, Cor. 5.5.14]). It is thus sufficient to prove this formula for constructible motivic sheaves. Using the fact that the Verdier duality functor $\mathbf{D}$ exchanges $*$'s and $!$'s as well as Theorem 3.4.1.10, we see that it is sufficient to prove the analogous formula obtained by considering functors of the form $(f \times g)^*$ and f^*, g^*, which is obvious.   □

Corollary 3.4.1.15 *Let X be a scheme together with $M, N \in DM_{h,lc}(X, \Lambda)$. If we denote by $\Delta : X \to X \times X$ the diagonal map, then*

$$\Delta^!(\mathbf{D}M \boxtimes N) \cong Hom(M, N) .$$

We have indeed:

$$\Delta^!(DM \boxtimes N) \cong D\Delta^*D(DM \boxtimes N)$$
$$\cong D\Delta^*(DDM \boxtimes DN)$$
$$\cong D(M \otimes DN)$$
$$\cong Hom(M, N).$$

3.4.2 Grothendieck-Lefschetz Formula

As in the previous paragraph, we assume that a ground field k is given, and all schemes are assumed to be separated of finite type over k.

Definition 3.4.2.1 Let X and Y be schemes, together with $M \in DM_{h,lc}(X, \Lambda)$ and $N \in DM_{h,lc}(Y, \Lambda)$. A *cohomological correspondence* from (X, M) to (Y, N) is a triple of the form (C, c, α), where (C, c) determines the commutative diagram

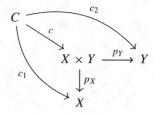

together with a map $\alpha : c_1^* M \to c_2^! N$ in $DM_h(C, \Lambda)$.

Remark 3.4.2.2 We have:

$$Hom(c_1^* M, c_2^! N) \cong Hom(c^* p_X^* M, c^! p_Y^! N) \cong c^! Hom(p_X^* M, p_Y^! N) \cong c^!(DM \boxtimes N).$$

Therefore, one can see α as a map of the form

$$\alpha : \Lambda \to c^!(DM \boxtimes N).$$

Remark 3.4.2.3 In the case where c_2 is proper, a cohomological correspondence induces a morphism in cohomology as follows. Let $a : X \to Spec(k)$ and $b : Y \to Spec(k)$ be the structural maps. We e have $ac_1 = bc_2$ and a co-unit map $(c_2)_* c_2^!(N) \to N$, whence a map:

$$a_* M \to a_*(c_1)_* c_1^* M \xrightarrow{a_*(c_1)_* \alpha} a_*(c_1)_* c_2^! N \cong b_*(c_2)_* c_2^! N \to b_* N.$$

In particular, one can consider the trace of such an induced map. By duality, in the case where c_1 is proper, we get an induced map in cohomology with compact support $b_! N \to a_! M$.

3.4.2.4 We observe that cohomological correspondences can be multiplied: given another cohomological correspondence (C', c', α') from (X', M') to (Y', N'), we define a new correspondence from $(X \times X', M \boxtimes M')$ to $(Y \times Y', N \boxtimes N')$ with

$$(C, c, \alpha) \otimes (C', c', \alpha') = (C \times C', c \times c', \alpha \boxtimes \alpha')$$

where $\alpha \boxtimes \alpha'$ is defined using the functoriality of the $\boxtimes$ operation together with the canonical Künneth isomorphisms seen in the previous paragraph:

$$\Lambda \cong \Lambda \boxtimes \Lambda \xrightarrow{\alpha \boxtimes \alpha'} c^!(\mathbf{D}M \boxtimes N) \boxtimes c'^!(\mathbf{D}M' \boxtimes N') \cong (c \times c')^!(\mathbf{D}(M \boxtimes M') \boxtimes (N \boxtimes N')) \, .$$

3.4.2.5 Correspondences can also be composed. Let (C, c, α) be a correspondence from (X, M) to (Y, N) as above, and let (D, d, β) be a correspondence from (Y, N) to (Z, P), with (D, d) corresponding to a commutative diagram of the form below, and $\beta : \Lambda \to d^!(\mathbf{D}N \boxtimes P)$ a map in in $DM_h(D, \Lambda)$.

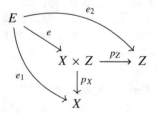

We form the following pullback square

$$
\begin{array}{ccc}
E & \xrightarrow{\lambda} & D \\
\downarrow{\mu} & & \downarrow{d_1} \\
C & \xrightarrow{c_2} & Y
\end{array}
$$

as well as the commutative diagram

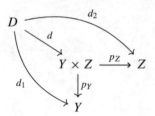

in which $e_1 = c_1\mu$ and $e_2 = d_2\lambda$. We then form $\alpha \boxtimes \beta$:

$$\Lambda \cong \Lambda \boxtimes \Lambda \xrightarrow{\alpha \boxtimes \beta} c^!(\mathbf{DM} \boxtimes N) \boxtimes d^!(\mathbf{DN} \boxtimes P) \cong (c \times d)^!((\mathbf{DM} \boxtimes N) \boxtimes (\mathbf{DN} \boxtimes P)).$$

Let $f = d_1\lambda = c_2\mu : E \to Y$ be the canonical map, and $\Delta : Y \to Y \times Y$ be the diagonal. We have the following Cartesian square

$$
\begin{array}{ccc}
E & \xrightarrow{(\mu,\lambda)} & C \times D \\
\varphi=(e_1,f,e_2)\downarrow & & \downarrow c \times d \\
X \times Y \times Z & \xrightarrow{1 \times \Delta \times 1} & X \times Y \times Y \times Z
\end{array}
$$

which induces an isomorphism (proper base change formula)

$$\varphi_!(\mu, \lambda)^* \cong (1 \times \Delta \times 1)^*(c \times d)_!.$$

In particular, it induces a canonical map

$$\kappa : (\mu, \lambda)^*(c \times d)^! \to \varphi^!(1 \times \Delta \times 1)^*$$

corresponding by adjunction to the composite

$$\varphi_!(\mu, \lambda)^*(c \times d)^! \cong (1 \times \Delta \times 1)^*(c \times d)_!(c \times d)^! \xrightarrow{\text{co-unit}} (1 \times \Delta \times 1)^*.$$

Let $\pi : X \times Y \times Z \to X \times Z$ be the canonical projection. There is a canonical map

$$\varepsilon : (1 \times \Delta \times 1)^*(\mathbf{DM} \boxtimes N \boxtimes \mathbf{DN} \boxtimes P) \to \pi^!(\mathbf{DM} \boxtimes P)$$

induced by the evaluation map

$$N \otimes \mathbf{DN} \to I_Y$$

together with the canonical identifications coming from appropriate Künneth formulas:

$$(1 \times \Delta \times 1)^*(\mathbf{DM} \boxtimes (N \boxtimes \mathbf{DN}) \boxtimes P) \cong \mathbf{DM} \boxtimes (N \otimes \mathbf{DN}) \boxtimes P$$

$$\mathbf{DM} \boxtimes I_Y \boxtimes P \cong \pi^!(\mathbf{DM} \boxtimes P).$$

We observe that $e = \pi\varphi$, so that $e^! \cong \varphi^!\pi^!$.

Definition 3.4.2.6 With the notations above, composing $(\mu, \lambda)^*(\alpha \boxtimes \beta)$ with the maps κ and ε defines the map

$$\beta \circ \alpha : \Lambda \cong (\mu, \lambda)^* \Lambda \to \varphi^! \pi^! (\mathbf{D}M \boxtimes P) \cong e^! (\mathbf{D}M \boxtimes P).$$

We define finally define the *composition of the correspondences* (C, c, α) and (D, d, β) as

$$(D, d, \beta) \circ (C, c, \alpha) = (E, e, \beta \circ \alpha).$$

3.4.2.7 This composition is only well defined up to isomorphism (since some choice of pull-back appears), but it is associative and unital up to isomorphism. The unit cohomological correspondence of (X, M) is given by

$$1_{(X,M)} = (X, \Delta, 1_M)$$

where $\Delta : X \to X \times X$ is the diagonal map and

$$1_M : \Lambda \to \Delta^! (\mathbf{D}M \boxtimes M) \cong Hom(M, M)$$

is the canonical unit map. In a suitable sense, this defines a symmetric monoidal bicategory, where the tensor product is defined as

$$(X, M) \otimes (Y, N) = (X \times Y, M \boxtimes N)$$

while the unit object if $(Spec(k), \Lambda)$.

3.4.2.8 To make this a little bit more precise, we must speak of the category of cohomological correspondences from (X, M) to (Y, N), in order to be able to express the fact that all the contructions and all the coherence isomorphisms (expressing the associativity and so on) are functorial. If (C, c, α) and (D, d, β) both are correspondences from (X, M) to (Y, N), a map

$$\sigma : (C, c, \alpha) \to (D, d, \beta)$$

is a pair $\sigma = (f, h)$, where $f : C \to D$ is a proper morphism such that $df = c$, while h is a homotopy

$$h : f_!(\alpha) \cong \beta$$

where $f_!(\alpha)$ is the map defined as

$$f_!(\alpha) : \Lambda \xrightarrow{\text{unit}} f_* \Lambda \xrightarrow{f_* \alpha} f_* c^! (\mathbf{D}M \boxtimes N) \cong f_* f^! d^! (\mathbf{D}M \boxtimes N) \xrightarrow{\text{co-unit}} d^! (\mathbf{D}M \boxtimes N).$$

This defines the symmetric monoidal bicategory $MCorr(k)$ whose objects are the pairs (X, M) formed of a k-scheme X equipped with a Λ-linear locally constructible h-motive M. In particular, for each pair of pairs (X, M) and (Y, N), there is the category of cohomological correspondences from (X, M) to (Y, N), denoted by $MCorr(X, M; Y, N)$ (in this paragraph, unless we make it explicit otherwise, we will only need the 1-category of such things, considering maps α as above in the homotopy category of h-motives).

Proposition 3.4.2.9 *All the objects of $MCorr(k)$ are dualizable. Moreover, the dual of a pair (X, M) is $(X, \mathbf{D}M)$.*

Proof Let (X, M), (Y, N) and (Z, P) be three objects of $MCorr(k)$. A cohomological correspondence from $(X \times Y, M \boxtimes N)$ to (Z, P) is determined by a morphism of k-schemes $c : C \to X \times Y \times Z$ together with a map

$$\alpha : \Lambda \to c^!(\mathbf{D}(M \boxtimes N) \boxtimes P).$$

A cohomological correspondence from (X, M) to $(Y \times Z, \mathbf{D}N \boxtimes P)$ is determined by a morphism of k-schemes $c : C \to X \times Y \times Z$ together with a map

$$\alpha : \Lambda \to c^!(\mathbf{D}M \boxtimes (\mathbf{D}N \boxtimes P)).$$

The Künneth formula

$$\mathbf{D}(M \boxtimes N) \boxtimes P \cong \mathbf{D}M \boxtimes (\mathbf{D}N \boxtimes P)$$

implies our assertion. $\square$

3.4.2.10 Let X be a scheme and M a locally constructible h-motive on X. We denote by $\Delta : X \to X \times X$ the diagonal map. There is a *transposed evaluation map*

$$ev^t_M : \mathbf{D}M \boxtimes M \to \Delta_* I_X$$

which corresponds by adjunction to the classical evaluation map

$$\Delta^*(\mathbf{D}M \boxtimes M) \cong Hom(M, I_X) \otimes M \to I_X.$$

Definition 3.4.2.11 Let (C, c, α) be a cohomological correspondence from (X, M) to (Y, N). In the case $(X, M) = (Y, N)$ we can form the following Cartesian square.

$$
\begin{array}{ccc}
F & \xrightarrow{\;\delta\;} & C \\
{\scriptstyle v}\downarrow & & \downarrow{\scriptstyle c} \\
X & \xrightarrow{\;\Delta\;} & X \times X
\end{array}
$$

The scheme F is called the *fixed locus* of the correspondence (C, c). The transposed evaluation map of M induces by proper base change a map

$$c^!(ev_M^t) : c^!(\mathbf{D}M \boxtimes M) \to c^!\Delta_* I_X \cong \delta_* p^! I_X \cong \delta_* I_F \,,$$

and thus, by adjunction, a map

$$ev_{M,c}^t : \delta^* c^!(\mathbf{D}M \boxtimes M) \to I_F \,.$$

The map $\alpha : \Lambda \to c^!(\mathbf{D}M \boxtimes M)$ finally induces a map

$$Tr(\alpha) : \Lambda \cong p^*\Lambda \to I_F$$

defined as the composition of $\delta^*\alpha$ with $ev_{M,c}^t$ (modulo the identification $\delta^*\Lambda \cong \Lambda$). The corresponding class

$$Tr(\alpha) \in H^0 \mathrm{Hom}_{DM_h(F,\Lambda)}(\Lambda, I_F)$$

is called the *characteristic class* of α.

Example 3.4.2.12 Let $f : X \to X$ be a morphism of schemes, and let M be a Λ-linear locally constructible h-motive on X, equipped with a map $\alpha : f^*M \to M$. Then $(X, (1_X, f), \mathbf{D}\alpha)$ is a cohomological correspondence from $(X, \mathbf{D}M)$ to itself, with

$$\mathbf{D}\alpha : 1_X^* \mathbf{D}M \cong \mathbf{D}M \to \mathbf{D}f^*M \cong f^! \mathbf{D}M \,.$$

If we form the Cartesian square

$$\begin{array}{ccc} F & \xrightarrow{\;\delta\;} & X \\ {\scriptstyle p}\downarrow & & \downarrow{\scriptstyle (1_X,f)} \\ X & \xrightarrow{\;\Delta\;} & X \times X \end{array}$$

we see that F is indeed the fixed locus of the morphism f. If $\Lambda \subset \mathbf{Q}$, then he associated characteristic class

$$Tr(\mathbf{D}\alpha) \in H^0 \mathrm{Hom}(\Lambda, I_F) \otimes \mathbf{Q} \cong CH_0(F) \otimes \mathbf{Q}$$

defines a 0-cycle on F (see Theorem 3.2.4.3). In the case where f only has isolated fixed points, we have

$$CH_0(F) \otimes \mathbf{Q} \cong CH_0(F_{red}) \otimes \mathbf{Q} \cong \oplus_{i \in I} CH_0(Spec(k_i)) \otimes \mathbf{Q}$$

where I is a finite set and each k_i is a finite field extension of k with $F_{red} = \coprod_i Spec(k_i)$. Using this decomposition, one can then express the characteristic class of α as a sum of local terms: the contributions of each summand $CH_0(Spec(k_i)) \otimes \mathbf{Q}$. For instance, if U is an open subset of X such that $f(U) \subset U$, and if $j : U \to X$ is the inclusion map, we can consider $M = j_! \Lambda$ and the canonical isomorphism $\alpha : f^* j_! \Lambda \to j_! \Lambda$, in which case $Tr(\alpha)$ is a way to count the number of fixed points of f in U with 'arithmetic multiplicities' (in the form of 0-cycles).

Remark 3.4.2.13 The notation $Tr(\alpha)$ is justified by Proposition 3.4.2.9: indeed, essentially by definition of the composition law for cohomological correspondences sketched in Paragraph 3.4.2.6, the characteristic class $Tr(\alpha)$ is the trace of the endomorphism (C, c, α) of the dualizable object (X, M). Indeed, the endomorphisms of $(Spec(k), \Lambda)$ in $MCorr(k)$ are determined by pairs (F, t) where F is a k-scheme and $t : \Lambda \to I_F$ is a section of the dualizing object of F in $DM_h(F, \Lambda)$.

Corollary 3.4.2.14 *For any cohomological correspondences (C, c, α) and (D, d, β) from (X, M) to itself, we have:*

$$Tr(\beta \circ \alpha) = Tr(\alpha \circ \beta) .$$

Corollary 3.4.2.15 *Let (C, c, α) be a cohomological correspondence from (X, M) to itself. If we see α as a map from $c_1^* M \to c_2^! M$, it determines a map*

$$\mathbf{D}\alpha : c_2^* \mathbf{D}M \cong \mathbf{D}c_2^! M \to \mathbf{D}c_1^* M \cong c_1^! \mathbf{D}M .$$

If $\tau : X \times X \to X \times X$ denotes the permutation of factors, the cohomological correspondence $(C, \tau c, \mathbf{D}\alpha)$ from $(X, \mathbf{D}M)$ to itself is the explicit description of the map obtained from (C, c, α) by duality. In particular:

$$Tr(\alpha) = Tr(\mathbf{D}\alpha) .$$

3.4.2.16 The formation of traces is functorial with respect to morphisms of correspondences. Let M be a locally constructible motive on a scheme X, and $f : C \to D$, $d : D \to X \times X$, $c = df$, be morphisms, with f proper. We form pull-back squares

$$p \left(\begin{array}{ccc} F & \xrightarrow{\delta} & C \\ \downarrow g & & \downarrow f \\ G & \xrightarrow{\varepsilon} & D \\ \downarrow q & & \downarrow d \\ X & \xrightarrow{\Delta} & X \times X \end{array} \right) c$$

and have a composition

$$f_*c^!(\mathbf{DM} \boxtimes N) \cong f_*f^!d^!(\mathbf{DM} \boxtimes N) \xrightarrow{\text{co-unit}} d^!(\mathbf{DM} \boxtimes N)$$

as well as a composition

$$f_*\delta_*I_F \cong \varepsilon_*g_*I_F \cong \varepsilon_*g_*g^!I_G \xrightarrow{\text{co-unit}} \varepsilon_*I_G.$$

One then checks right away that the following square commutes.

$$
\begin{array}{ccc}
f_*c^!(\mathbf{DM} \boxtimes N) & \longrightarrow & d^!(\mathbf{DM} \boxtimes N) \\
{\scriptstyle f_*c^!(ev_M^t)}\Big\downarrow & & \Big\downarrow {\scriptstyle g_*d^!(ev_M^t)} \\
f_*\delta_*I_F & \longrightarrow & \varepsilon_*I_G
\end{array}
$$

This implies immediately that, for any map $\alpha : \Lambda \to c^!(\mathbf{DM} \boxtimes M)$, we have:

$$Tr(\alpha) = Tr(f_!(\alpha)).$$

3.4.2.17 Proper maps act on cohomological correspondences as follows. We consider a proper morphism of geometric correspondences, by which we mean a commutative square of the form

$$
\begin{array}{ccc}
C & \xrightarrow{\varphi} & D \\
{\scriptstyle c=(c_1,c_2)}\Big\downarrow & & \Big\downarrow {\scriptstyle d=(d_1,d_2)} \\
X \times X' & \xrightarrow{f \times f'} & Y \times Y'
\end{array}
$$

in which $f : X \to Y$, $f' : X' \to Y'$ and $\varphi : C \to D$ are proper map, together with locally constructible h-motives M on X and M' on X'. Given a cohomological correspondence from (X, M) to (X', M') of the form (C, c, α), we have a cohomological correspondence from $(X, f_!M)$ to $(X', f'_!M')$

$$(f, f')_!(C, c, \alpha) = (C, d\varphi, (f, f')_!(\alpha))$$

defined as follows. If, furthermore, the commutative square above is Cartesian, the map $(f, f')_!(\alpha)$ is the induced map

$$\Lambda \xrightarrow{\text{unit}} \varphi_*\Lambda \xrightarrow{\varphi_*\alpha} \varphi_*c^!(\mathbf{DM} \boxtimes M') \cong d^!(f \times f')_*(\mathbf{DM} \boxtimes M') \cong d^!(\mathbf{D}f_!M \boxtimes f'_!M')$$

Otherwise, we consider the induced proper map

$$g : C \to E = X \times X' \times_{Y \times Y'} D$$

and apply the preceding construction to $g_!(\alpha)$, replacing C by E.

In the case where $(X, M) = (X', M')$ and $f = f'$, we simply write

$$f_!(\alpha) = (f, f)_!(\alpha).$$

Theorem 3.4.2.18 (Lefschetz-Verdier Formula) *We consider a commutative square of k-schemes of finite type of the form*

$$
\begin{array}{ccc}
C & \xrightarrow{\;\varphi\;} & D \\
{\scriptstyle c=(c_1,c_2)}\downarrow & & \downarrow{\scriptstyle d=(d_1,d_2)} \\
X \times X & \xrightarrow{\;f \times f\;} & Y \times Y
\end{array}
$$

in which both f and φ are proper, as well as a locally constructible h-motive M on X, together with a map $\alpha : \Lambda \to c^!(\mathbf{D}M \boxtimes M)$. Let F and G be the fixed locus of (C, c) and (D, d) respectively. Then the induced map $\psi : F \to G$ is also proper, and

$$\psi_!(Tr(\alpha)) = Tr(f_!(\alpha)).$$

Proof The functoriality of the trace explained in 3.4.2.16 shows that it is sufficient to prove the theorem in the case where the square is Cartesian. We check that the two maps

$$(f \times f)_*(\mathbf{D}M \boxtimes M) \cong (\mathbf{D}f_!M \boxtimes f_!M) \xrightarrow{\;ev^t_{f_!M}\;} \Delta_* I_Y$$

and

$$(f \times f)_*(\mathbf{D}M \boxtimes M) \xrightarrow{\;(f \times f)_*(ev^t_M)\;} (f \times f)_* \Delta_* I_X \cong \Delta_* f_* I_X \cong \Delta_* f_! f^! I_Y \xrightarrow{\;\text{co-unit}\;} \Delta_* I_Y$$

are equal (where we have denoted by the same symbol the diagonal of X and the diagonal of Y). By duality, this amounts to check that the unit map

$$\Delta_* \Lambda \to M \boxtimes \mathbf{D}M$$

is compatible with the push-forward f_*. This is a fancy way to say that f_*M has a natural $f_*\Lambda$-algebra structure, which comes from the fact that the functor f^* is symmetric monoidal. The Lefschetz-Verdier Formula follows then right away. $\quad\square$

Remark 3.4.2.19 When $\Lambda = \mathbf{Q}$, the operator $\psi_!$ coincides with the usual push forward of 0-cycles: seen as a map

$$\psi_! : H^0\mathrm{Hom}(\Lambda, I_F) \to H^0\mathrm{Hom}(\Lambda, I_G).$$

Theorem 3.4.2.20 (Additivity of Traces) *Let* $c = (c_1, c_2) : C \to X \times X$ *be a correspondence of k-schemes. We consider a cofiber sequence*

$$M' \to M \to M''$$

in $DM_{h,lc}(X)$ *as well as maps*

$$\alpha' : c_1^* M' \to c_2^! M', \ \alpha : c_1^* M \to c_2^! M, \ \alpha'' : c_1^* M'' \to c_2^! M''$$

in $DM_{h,lc}(C)$ *so that the diagram below commutes (in the sense of* ∞*-categories).*

$$
\begin{array}{ccc}
c_1^* M' \longrightarrow c_1^* M \longrightarrow c_1^* M'' \\
\downarrow{\scriptstyle\alpha'} \qquad\quad \downarrow{\scriptstyle\alpha} \qquad\quad \downarrow{\scriptstyle\alpha''} \\
c_2^! M' \longrightarrow c_2^! M \longrightarrow c_2^! M''
\end{array}
$$

Then the following formula holds.

$$Tr(\alpha) = Tr(\alpha') + Tr(\alpha'')$$

The proof is given in the paper of Jin and Yang [27, Theorem 4.2.8] using the language of algebraic derivators, which is sufficient for our purpose (note however that, by Balzin's work [6, Theorem 2], it is clear that one can go back and forth between the language of fibred ∞-categories and the one of algebraic derivators). The additivity of traces can be extended to more general homotopy colimits; see Gallauer's thesis [21].

Remark 3.4.2.21 It is pleasant to observe that, when $\Lambda = \mathbf{Q}$, this is the classical push-forward of 0-cycles. Lefschetz-Verdier Formula is particularly relevant in the case where $Y = Spec(k)$, and F consists of isolated points in X (in which case it is called the Grothendieck-Lefschetz formula). Indeed, $f_!(M)$ is then the cohomology of X with compact support with coefficients in M, so that $Tr(f_!(\alpha))$ is the ordinary trace of the endomorphism $f_!(\alpha) : f_!(M) \to f_!(M)$, which can be computed through ℓ-adic realizations as an alternating sum of ordinary traces of linear maps. On the other hand, $\psi_!(Tr(\alpha))$ is the sum of traces of the endomorphisms induced by α on each $p_! x^* M$, where x runs over the points of F, with $p : Spec(\kappa(x)) \to Spec(k)$ the structural map. In the particular case discussed at the end of Example 3.4.2.12, this shows that one can compute the number of fixed points with geometric multiplicities of a endomorphism of a k-scheme $f : X \to X$ with isolated fixed points which extends to an endomorphism of a compactification of X and whose graph is transverse to the diagonal, using the trace of the induced endomorphism of the motive with compact support of X. For the Frobenius map, such an extension is automatic, so that We can count rational points of any separated

$\mathbf{F}_q$-scheme of finite type X_0 over a finite field $\mathbf{F}_q$ with the Grothendieck-Lefschetz formula

$$\#X(\mathbf{F}_q) = \sum_i (-1)^i \operatorname{Tr}\big(F : H_c^i(X, \mathbf{Q}_\ell) \to H_c^i(X, \mathbf{Q}_\ell)\big),$$

where X is the pull-back of X_0 on the algebraic closure $\mathbf{F}_q$, and where F is a the map induced by the geometric Frobenius (i.e. where one considers the correspondence defined by the transposed graph of the arithmetic Frobenius). Indeed, using the additivity of traces, it is in fact sufficient to prove this formula in the case where X is smooth and projective, in which case the classical Lefschetz formula applies. We will now prove a more general version of it: we will consider arbitrary (locally) constructible motivic sheaves as coefficients.

3.4.2.22 Let p be a prime number, $r > 0$ a natural number, and $q = p^r$. Let $k_0 = \mathbf{F}_q$ be the finite field with q elements, and let us choose an algebraic closure k of k_0. Given a $\mathbf{F}_p$-scheme X, we denote by

$$F_X : X \to X$$

the *absolute Frobenius* of X, given by the identity on the underlying topological space, and by $a \mapsto a^p$ on the structural sheaf $\mathcal{O}_X$. The absolute Frobenius is a natural transformation from the identity of the category of $\mathbf{F}_p$-schemes to itself. In particular, for any morphism of k-schemes $u : U \to X$, there is a commutative square

$$\begin{array}{ccc} U & \xrightarrow{\ F_U\ } & U \\ {\scriptstyle u}\downarrow & & \downarrow{\scriptstyle u} \\ X & \xrightarrow{\ F_X\ } & X \end{array}$$

and thus a comparison map:

$$F_{U/X} = (u, F_U) \colon U \to F_X^{-1}(U) = X \times_X U$$

called the *relative Frobenius* of U over X. In case X_0 is a k_0-scheme, the rth iteration of the absolute Frobenius

$$F_{X_0}^r : X_0 \to X_0$$

is often called the *q-absolute Frobenius of X_0* (and has the feature of being a map of k_0-schemes). By base change to k, it induces the *geometric Frobenius of X*, i.e. the morphism of k-schemes

$$\phi_r : X \to X,$$

where $X = \mathrm{Spec}(k) \times_{\mathrm{Spec}(k_0)} X_0$. Following Deligne's conventions, sheaves (or motives) on X_0 will often be denoted by M_0, and the pullback of M_0 along the canonical projection $X \to X_0$ will be written M. The map $k \to k$, defined by $x \mapsto x^q$ is an automorphism of k_0-algebras, which induces an isomorphism of k_0-schemes

$$Frob_q : \mathrm{Spec}(k) \to \mathrm{Spec}(k).$$

It induces an isomorphism of k_0-schemes

$$Frob_{q,X} = (Frob_q \times_{\mathrm{Spec}(k_0)} 1_{X_0}) : X \to X$$

whose composition with ϕ_r is nothing else than the absolute Frobenius of X. The map

$$Frob_{q,X}^{-1} : X \to X$$

is often called the *arithmetic Frobenius of X*.

Lemma 3.4.2.23 *Let X be a locally noetherian $\mathbf{F}_p$-scheme. The functor*

$$F_X^* : DM_h(X, \Lambda) \to DM_h(X, \Lambda)$$

is the identity.

Proof Let $a : X \to \mathrm{Spec}(\mathbf{F}_p)$ be the structural map. We have a commutative diagram of the form

$$
\begin{array}{ccc}
X & \xrightarrow{\;F_X\;} & X \\
\downarrow{\scriptstyle a} & & \downarrow{\scriptstyle a} \\
\mathrm{Spec}(\mathbf{F}_p) & =\!=\!= & \mathrm{Spec}(\mathbf{F}_p)
\end{array}
$$

in which the map F_X is a universal homeomorphism (being integral, radicial and surjective) and thus invertible locally for the h-topology. In other words, the square above is Cartesian locally for the h-topology. By h-descent, the functor

$$F_X^* : DM_h(X, \Lambda) \to DM_h(X, \Lambda)$$

thus acts as the identity. $\qquad\square$

Remark 3.4.2.24 For a k_0-scheme X_0, since the composition of the geometric Frobenius $\phi_r : X \to X$ with the inverse of the arithmetic Frobenius is the absolute Frobenius, this shows that considering actions of the geometric Frobenius or of the arithmetic Frobenius amount to the same thing, at least as far as motivic sheaves are concerned. In fact the previous lemma is also a way to define such actions.

Let M_0 be a motivic sheaf on X_0, i.e. an object of $DM_h(X_0, \Lambda)$. Since $F_X = Frob_{q,X} \phi_r$ we have

$$F_X^*(M) = M \simeq \phi_r^* \, Frob_{q,X}^*(M) \, .$$

On the other hand, since M_0 is defined over k_0, and $M = a^*(M_0)$, there is a canonical isomorphism

$$Frob_{q,X}^*(M) \cong M \, .$$

Therefore, we have a canonical isomorphism

$$\phi_r^*(M) \cong M = (1_X)^!(M) \, .$$

Since the locus of fixed points of ϕ_r is precisely the (finite) set $X(k_0)$ of rational points of X_0 (seen as a discrete algebraic variety over k), the Verdier trace of the isomorphism above defines a class

$$L(M_0) = Tr\left(\phi_r^*(M) \overset{\cong}{\to} (1_X)^!(M) \right) \in H_0(X(k_0), \Lambda) \, .$$

If p is invertible in Λ, we have simply

$$H_0(X(k_0), \Lambda) \cong \Lambda^{X(k_0)} \, .$$

Under this identification, the obvious function

$$\Lambda^{X(k_0)} \to \Lambda \, , \quad f \mapsto \int f = \sum_{x \in X(k_0)} f(x)$$

coincides with the operator

$$\psi_! : H_0(X(k_0), \Lambda) \to H_0(\mathrm{Spec}(k), \Lambda) = \Lambda$$

induced by the structural map $\psi : X(k_0) \to \mathrm{Spec}(k)$.

The action of Frobenius is functorial: for any map $f : X_0 \to Y_0$, the induced action of ϕ_r^* on $f_!(M)$ via the proper base change isomorphism

$$\phi_r^* f_! \cong f_! \phi_r^*$$

coincides with the action defined as above in the case of $f_!(M_0)$. There is a similar compatibility with the canonical isomorphism $\phi_r^* f^* \cong f^* \phi_r^*$.

If X_0 is proper and if $j : U_0 \to X_0$ is an open immersion, for any M_0 locally constructible in $DM_h(U_0, \mathbf{Q})$, we thus get, as a special case of the Lefschetz-Verdier Formula (Theorem 3.4.2.18)

$$Tr\Big(\phi_r^* : a_!(M) \to a_!(M)\Big) = \int L(j_! M_0) \in \mathbf{Q}$$

where $a : U \to \mathrm{Spec}(k)$ is the structural morphism.

Theorem 3.4.2.25 (Grothendieck-Lefschetz Formula) *Let $j : U_0 \to X_0$ is an open immersion into a proper scheme of finite type over a finite field k_0 and let M_0 be a locally constructible motivic sheaf in $DM_h(U_0, \mathbf{Q})$. For each rational point x of U_0, we denote by M_x the fiber of M at the induced geometric point of U, on which there is a canonical action of the geometric Frobenius (as a particular case of the construction of Remark 3.4.2.24). Then*

$$Tr\Big(\phi_r^* : a_!(M) \to a_!(M)\Big) = \sum_{x \in U(k_0)} Tr\Big(\phi_r^* : M_x \to M_x\Big).$$

Proof The case where $M_0 = \mathbf{Q}$ is constant is well known (see Remark 3.4.2.21). This proves the case where $M_0 = p_!(\mathbf{Q})$ for a map $p : Y_0 \to U_0$. The case of a direct factor of $p_!(\mathbf{Q})$ with Y_0 smooth and projective can be proved in the same way: the projector defining our motive is then given by some $\dim(Y)$-dimensional cycle α on $Y \times Y$ supported on $Y \times_X Y$ (see Theorem 3.2.4.3). We then observe that the Grothendieck-Lefschetz fixed point formula holds (using proper base change formula and Olsson's computation of local terms [34, Prop. 5.5]). On the other hand, we see that the shift $[i]$ and the Tate twist (n) are compatible with traces (they consist in multiplying by $(-1)^i$ and by 1, respectively). By the additivity of traces, we are comparing two numbers which only depend on the class of M_0 in the Grothendieck group $K_0(DM_{h,lc}(U_0, \mathbf{Q}))$, and it is sufficient to consider the case where $U_0 = X_0$ is projective. Using Bondarko's theory of motivic weights [9, Prop. 3.3], we see that any class in $K_0(DM_{h,lc}(U_0, \mathbf{Q}))$ is a linear combination of classes of motives which are direct factors of $p_*(\mathbf{Q})(n)[i]$ for $n, i \in \mathbf{Z}$ and $p : Y_0 \to U_0$ a projective morphism, with Y_0 smooth and projective. This proves the formula in general. $\square$

Remark 3.4.2.26 When $M_0 = p_!(\mathbf{Q})$, with $p : Y_0 \to U_0$ separated of finite type, the Grothendieck-Lefschetz Formula expresses the trace of the Frobenius action on cohomology with compact support of Y as a sum of the traces of the action of Frobenius on cohomology with compact support of the fibers Y_x of Y over each k_0-rational point x of U. One can do similar constructions replacing the geometric Frobenius action by any (functorially given) automorphism of k-schemes, such as the identity. The computation of the local terms given by the Lefschetz-Verdier Trace Formula can then be rather involved. For instance, in the case of the identity (which means that we want to compute Euler-Poincaré characteristic of cohomology with compact support), the naive formula tends to fail (at least in positive characteristic). The Grothendieck-Ogg-Shafarevitch Formula is such a nontrivial computation in the case where M_0 is dualizable on a smooth curve U_0: it measures the defect of the naive formula in terms of Swan conductors.

Acknowledgments These notes are an account of a series of lectures I gave at the LMS-CMI Research School 'Homotopy Theory and Arithmetic Geometry: Motivic and Diophantine Aspects', in July 2018, at the Imperial College London. I am grateful to Shachar Carmeli for having allowed me to use the notes he typed from my lectures, and to Kévin François for finding a gap in the proof of the motivic generic base formula. While preparing these lectures and writing these notes, I was partially supported by the SFB 1085 "Higher Invariants" funded by the Deutsche Forschungsgemeinschaft (DFG).

References

1. G. Almkvist, K-theory of endomorphisms. J. Algebra **55**(2), 308–340 (1978)
2. B. Antieau, D. Gepner, J. Heller, K-theoretic obstructions to bounded t-structures. Inventiones Math. **216**(1), 241–300 (2019)
3. M. Artin, A. Grothendieck, J.-L. Verdier, Théorie des topos et cohomologie étale des schémas, in *Lecture Notes in Mathematics*, vol. 270, 305, 569 (Springer, Berlin, 1972–1973). Séminaire de Géométrie Algébrique du Bois-Marie 1963–1964 (SGA 4), Dirigé par M. Artin, A. Grothendieck et J.-L. Verdier. Avec la collaboration de N. Bourbaki, P. Deligne et B. Saint-Donat.
4. J. Ayoub, Six opérations de Grothendieck et cycles évanescents dans le monde motivique I. Astérisque **314**, 476 (2007)
5. J. Ayoub, La réalisation étale et les opérations de Grothendieck. Ann. Sci. École Norm. Sup. **47**(1), 1–145 (2014)
6. E. Balzin, Reedy model structures in families (2019). arXiv:1803.00681v2
7. A. Blanc, M. Robalo, B. Toën, G. Vezzosi, Motivic realizations of singularity categories and vanishing cycles. J. Éc. Polytech. Math. **5**, 651–747 (2018)
8. A.J. Blumberg, D. Gepner, G. Tabuada, A universal characterization of higher algebraic K-theory. Geom. Topol. **17**(2), 733–838 (2013)
9. M.V. Bondarko, Weights for relative motives: relation with mixed complexes of sheaves. Int. Math. Res. Not. IMRN **17**, 4715–4767 (2014)
10. D.-C. Cisinski, Descente par éclatements en K-théorie invariante par homotopie. Ann. Math. **177**(2), 425–448 (2013)
11. D.-C. Cisinski, Higher categories and homotopical algebra, in *Cambridge Studies in Advanced Mathematics*, vol. 180 (Cambridge University, Cambridge, 2019)
12. D.-C. Cisinski, F. Déglise, Integral mixed motives in equal characteristic. Documenta Math.. Extra Volume: Alexander S. Merkurjev's Sixtieth Birthday (2015), pp. 145–194
13. D.-C. Cisinski, F. Déglise, Étales motives. Compositio Math. **152**, 556–666 (2016)
14. D.-C. Cisinski, F. Déglise, Triangulated categories of mixed motives, in *Springer Monographs in Mathematics* (Springer, New York, 2019)
15. D.-C. Cisinski, G. Tabuada, Non-connective K-theory via universal invariants. Compos. Math. **147**(4), 1281–1320 (2011)
16. P. Deligne, La conjecture de Weil I. Publ. Math. IHES **43**, 273–307 (1974)
17. P. Deligne, Cohomologie étale SGA 4 1/2, in *Lecture Notes in Mathematics*, vol. 569 (Springer, Berlin, 1977)
18. P. Deligne, La conjecture de Weil II. Publ. Math. IHES **52**, 137–252 (1980)
19. K. Fujiwara, Rigid geometry, Lefschetz-Verdier trace formula and Deligne's conjecture. Invent. Math. **127**(3), 489–533 (1997)

20. K. Fujiwara, A proof of the absolute purity conjecture (after Gabber), in *Algebraic geometry 2000, Azumino (Hotaka)*. Advanced Studies in Pure Mathematics, vol. 36 (Mathematical Society, Japan, 2002), pp. 153–183
21. M. Gallauer, Traces in monoidal derivators, and homotopy colimits. Adv. Math. **261**, 26–84 (2014)
22. A. Grothendieck, Cohomologie ℓ-adique et fonctions L, in *Lecture Notes in Mathematics*, vol. 589 (Springer, Berlin, 1977). Séminaire de Géométrie Algébrique du Bois–Marie 1965–1966 (SGA 5)
23. A. Grothendieck, Revêtements étales et groupe fondamental, in *Documents Mathématiques*, vol. 3 (Society Mathematical France, Paris, 2003). Séminaire de Géométrie Algébrique du Bois–Marie 1960–1961 (SGA 1). Édition recomposée et annotée du LNM 224, Springer, 1971
24. M. Hoyois, The six operations in equivariant motivic homotopy theory. Adv. Math. **305**, 197–279 (2017)
25. L. Illusie, Miscellany on traces in ℓ-adic cohomology: a survey. Jpn. J. Math. **1**, 107–136 (2006)
26. L. Illusie, Y. Laszlo, F. Orgogozo (eds.), *Travaux de Gabber sur l'uniformisation locale et la cohomologie étale des schémas quasi-excellents*. (Société Mathématique de France, Paris, 2014). Séminaire à l'École Polytechnique 2006–2008. Avec la collaboration de Frédéric Déglise, Alban Moreau, Vincent Pilloni, Michel Raynaud, Joël Riou, Benoît Stroh, Michael Temkin and Weizhe Zheng, Astérisque No. 363–364
27. F. Jin, E. Yang, Künneth formulas for motives and additivity of traces (2018). arXiv:1812.06441
28. S. Kelly, Voevodsky motives and ℓdh-descent. Astérisque **391**, iv+125 (2017)
29. Y. Liu, W. Zheng, Gluing restricted nerves of ∞-categories (2015). arXiv:1211.5294v4
30. Y. Liu, W. Zheng, Enhanced six operations and base change theorem for higher Artin stacks (2017). arXiv:1211.5948v3
31. J. Lurie, Higher topos theory, in *Annals of Mathematics Studies*, vol. 170 (Princeton University, Princeton, 2009)
32. J. Lurie, *Higher Algebra* (2017). https://www.math.ias.edu/~lurie/papers/HA.pdf
33. N. Naumann, Algebraic independence in the Grothendieck ring of varieties. Trans. Am. Math. Soc. **359**(4), 1653–1683 (2007)
34. M. Olsson, Borel-Moore homology, Riemann-Roch transformations, and local terms. Adv. Math. **273**, 56–123 (2015)
35. M. Olsson, Motivic cohomology, localized Chern classes, and local terms. Manuscripta Math. **149**(1–2), 1–43 (2016)
36. R. Pink, On the calculation of local terms in the Lefschetz-Verdier trace formula and its application to a conjecture of Deligne. Ann. Math. **135**(3), 483–525 (1992)
37. N. Ramachandran, Zeta functions, Grothendieck groups, and the Witt ring. Bull. Sci. Math. **139**(6), 599–627 (2015)
38. M. Robalo, *Motivic Homotopy Theory of Non-commutative Spaces*, PhD thesis (2014). https://webusers.imj-prg.fr/~marco.robalo/these.pdf
39. M. Robalo, K-theory and the bridge from motives to noncommutative motives. Adv. Math. **269**, 399–550 (2015)
40. O. Röndigs, *Functoriality in Motivic Homotopy Theory* (2005). https://www.math.uni-bielefeld.de/~oroendig/functoriality.dvi
41. D. Rydh, Submersions and effective descent of étale morphisms. Bull. Soc. Math. France **138**(2), 181–230 (2010)
42. M. Schlichting, Negative K-theory of derived categories. Math. Z. **253**(1), 97–134 (2006)
43. A. Suslin, V. Voevodsky, Singular homology of abstract algebraic varieties. Invent. Math. **123**(1), 61–94 (1996)
44. Y. Varshavsky, A proof of a generalization of Deligne's conjecture. Electron. Res. Announc. Am. Math. Soc. **11**, 78–88 (2005)
45. Y. Varshavsky, Lefschetz-Verdier trace formula and a generalization of a theorem of Fujiwara. Geom. Funct. Anal. **17**(1), 271–319 (2007)

46. Y. Varshavsky, Intersection of a correspondence with a graph of Frobenius (2015). arXiv:1405.6381v3
47. V. Voevodsky, Homology of schemes. Selecta Math. (N.S.) 2(1), 111–153 (1996)
48. V. Voevodsky, A^1-homotopy theory, in *Proceedings of the International Congress of Mathematicians (Berlin, 1998)*, vol. I (1998), pp. 579–604
49. V. Voevodsky, A. Suslin, E.M. Friedlander, Cycles, transfers and motivic homology theories, in *Annals of Mathematics Studies*, vol. 143 (Princeton University, Princeton, 2000)

Chapter 4
Étale Homotopy and Obstructions to Rational Points

Tomer M. Schlank

Abstract These notes are supposed to serve as a condensed but approachable guide to the way étale homotopy can be used to study rational points. I hope readers from different backgrounds will find it useful, but it is probably most suitable for a reader with some background in algebraic geometry who is not necessarily as familiar with modern homotopical and ∞-categorical methods. The original definition of the étale homotopy type is due to Artin and Mazur, and the idea was further developed by Friedlander. In recent years there has been a lot of activity around étale homotopy and its applications.

4.1 Introduction

A natural question in commutative algebra is the following:

Question Let R be a commutative ring and let

$$\begin{cases} f_1(x_1, \ldots, x_n) = 0 \\ f_2(x_1, \ldots, x_n) = 0 \\ \vdots \\ f_m(x_1, \ldots, x_n) = 0 \end{cases} \qquad (*)$$

be a system of polynomial equations over R. Is there a solution over R?

T. M. Schlank (✉)
Einstein Institute of Mathematics, The Hebrew University of Jerusalem Givat Ram, Jerusalem, Israel
e-mail: tomer.schlank@mail.huji.ac.il

© The Author(s), under exclusive license to Springer Nature Switzerland AG 2021   107
F. Neumann, A. Pál (eds.), *Homotopy Theory and Arithmetic Geometry – Motivic and Diophantine Aspects*, Lecture Notes in Mathematics 2292,
https://doi.org/10.1007/978-3-030-78977-0_4

Algebraic geometry teaches us that we should think of this question "geometrically". Namely, to R one can associate a scheme Spec (R), which is a certain kind of a "space" and similarly to the ring

$$A = R[x_1, \ldots, x_n] / (f_1, \ldots, f_m)$$

we can associate the scheme Spec (A). The canonical map of rings $g: R \to A$ corresponds to a map of schemes $f = \text{Spec}(g): \text{Spec}(A) \to \text{Spec}(R)$ and a *solution* to $(*)$ corresponds precisely to a *section* of the map f (or equivalently a retraction of g). In particular, the system of equations $(*)$ has a solution if and only if f has a section. Generalizing from affine schemes to general schemes, we are interested in a method that given a map $f: X \to Y$ of schemes, would determine (or at least give some conditions for) whether f admits a section.

This is a very general type of a question that one can contemplate also for other kinds of "spaces". For example, when $f: X \to Y$ is a map of *topological spaces*. As a first approximation to this question, one can ask whether f admits a section *up to homotopy*. Put somewhat differently, we can discard some of the information and think of f as a map between the *homotopy types* of X and Y and ask whether this map of homotopy types admits a section.

The advantage in passing from topological spaces to homotopy types, is that the question can now be approached by the tools of algebraic topology. More concretely, one can use *obstruction theory*. This is a method to produce a sequence of "obstructions" to the existence of a section (up to homotopy), which are elements $o_n \in H^{n+1}(X, \pi_n(F))$, where F is the homotopy fiber of f. The vanishing of all the o_n-s is equivalent to the existence of a section up to **homotopy**. The main point being that those elements o_n are "algebraic" and thus fairly computable in many interesting situations. While their vanishing can not guarantee the existence of an actual section of f in the category of **topological spaces**, the *non-vanishing* will prove that such a section does not exist.

To apply these techniques to algebraic geometry, we need a way to transport the problem from the world of schemes to the world of topological spaces, so we can further push it to the world of homotopy types and use obstruction theory. The naive option to use just the underlying (Zariski) topological spaces of X and Y does not produce anything interesting as it is very coarse. In some special cases though, there is a suitable variant. For example, if $f: X \to Y$ is a map of algebraic varieties over $\mathbb{C}$, we can consider the sets of $\mathbb{C}$-points $X(\mathbb{C})$ and $Y(\mathbb{C})$ with the topology inherited from the usual one on $\mathbb{C}$. This captures much more of the "geometry" of the situation. The map f induces a continuous map $f^{an}: X(\mathbb{C}) \to Y(\mathbb{C})$ and of course any section of f would also induce a section of f^{an}. Thus, any obstruction to an existence of a section of f^{an} would also constitute an obstruction for the existence of a section of f (though perhaps a less strict one).

When we work with more general schemes, like algebraic varieties over finite fields, we do not have a natural way to construct an honest topological space. Nevertheless, using the *étale topology* we can construct from each scheme X a certain *étale topos* $X_{\acute{e}t}$ in a functorial way. A topos is a categorical gadget which generalizes the notion of a topological space by abstracting the properties of the category of sheaves of sets on a topological space and using them as axioms. For an algebraic variety X over $\mathbb{C}$, the étale topos $X_{\acute{e}t}$ carries a big part of the information in the actual topological space $X(\mathbb{C})$. From any topos $\mathcal{X}$, one can further construct a homotopy type $|\mathcal{X}|$ and then proceed using obstruction theory as before (to be precise, $|\mathcal{X}|$ is rather a "pro-homotopy type", we shall return to this point later). Thus, we divide the problem of finding a section into two steps:

$$\text{Scheme} \rightarrow \text{Topos} \rightarrow \text{Homotopy Type,}$$

which represent the passage from algebraic geometry to "topology" and then to homotopy theory.

One can also develop a variant of the above theory to the "relative situation". Instead of having one map $f: X \rightarrow Y$, say, of topological spaces, we can consider a commutative diagram

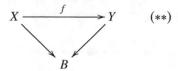

$$(**)$$

where B is some "base space". Over each point $b \in B$, the map f restricts to a map of fibers $f_b: X_b \rightarrow Y_b$. Hence, we can think of $X \rightarrow B$ and $Y \rightarrow B$ as families of spaces parameterized by B and f as a family of maps varying continuously over B. A section of f, commuting with the projection to B, would then be a continuous family of sections. There are now several ways to "homotopize" the situation. It turns out that it is possible (and profitable) to keep B being an actual topological space, and turn only the fibers X_b and Y_b into homotopy types. The corresponding obstruction theory would then invlove sheaves of groups over B, but other than that, it is quite analogous. Furthermore, if $(**)$ is a diagram of schemes, one can similarly produce sheaves of *relative* homotopy types $|X_{\acute{e}t}/B_{\acute{e}t}|$ and $|Y_{\acute{e}t}/B_{\acute{e}t}|$ over the topos $B_{\acute{e}t}$ and apply the said obstruction theory. We note that the relative theory has non-trivial applications even for the absolute case. Indeed, the absolute case $f: X \rightarrow Y$ can be treated as a special case of the relative one by taking $B = \text{pt}$. But there is also another option. We can take $B = Y$ with $f: X \rightarrow Y$ and $\text{Id}: Y \rightarrow Y$ as the structure maps. The second option produces a more refined obstruction theory as it remembers the topos $Y_{\acute{e}t}$ rather than merely its homotopy type.

The relative theory requires a good notion of a "sheaf of homotopy types". This notion is highly non-trivial as maps of homotopy types are on the one hand defined only up to homotopy, but on the other, for the obstruction theory to work, these homotopies need to be chosen in a certain "coherent way". The correct notion turns out to be that of an ∞-topos, which is itself a special case of an ∞-category. The plan of this course is thus as follows: We shall begin by reviewing the basics of ∞-categories. We proceed to discuss ∞-topoi and their associated (relative) homotopy type. We then review the classical obstruction theory for sections of maps of homotopy type. We proceed with explaining the theory of étale topoi which produces ∞-topoi from schemes and how the general theory specializes to this case. Finally, we show some explicit applications of this theory to rational points and zero-cycles on varieties and to embedding problems in Galois theory.

4.2 ∞-Categories

4.2.1 *Motivation*

The first approximation to an ∞-category is a category enriched in homotopy types. That is, like a category, but such that for each pair of objects $X, Y \in C$ we have a topological space of morphisms $\mathrm{Map}_C(X, Y)$ defined up to homotopy equivalence and the composition law is continuous and associative up to homotopy. The reason that this is only the first approximation can be explained as follows. Let

$$ X \xrightarrow{f} Y \xrightarrow{g} Z \xrightarrow{h} W \xrightarrow{k} U $$

be a 4-tuple of composable morphisms. Suppose we have chosen specific homotopies witnessing the associativity of all the possible compositions (called *associators*). From this we can produce a diagram of homotopies called the "Stasheff pentagon":

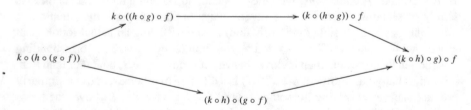

This diagram represents a map

$$ P_{f,g,h,k} \colon S^1 \to \mathrm{Map}(X, U). $$

We would like this map to be *null-homotopic*, meaning that we have chosen our associators "coherently". The data of such a contraction is an extension of $P_{f,g,h,k}$ to a map

$$ \tilde{P}_{f,g,h,k} \colon D^2 \to \mathrm{Map}(X, U). $$

An ∞-category carries this additional data for all 4-tuples (f, g, h, k) as above. This is only the tip of an infinite iceberg as these coherence homotopies $\tilde{P}_{f,g,h,k}$

themselves (*coherentors*) need to satisfy higher coherence conditions. That is, given a 5-tuple of composable morphisms

$$X \xrightarrow{f} Y \xrightarrow{g} Z \xrightarrow{h} W \xrightarrow{k} U \xrightarrow{l} V$$

we can use the different associators and coherentors to build a map

$$Q_{f,g,h,k,l} \colon S^2 \to \mathrm{Map}\,(X, U)\,.$$

Our ∞-category will also carry for every 5-tuple of composable morphisms the data of an extension of $Q_{f,g,h,k,l}$ from S^2 to D^3. This continues ad infinitum, so it is not surprising that finding a precise and workable formalisation of ∞-categories was a non-trivial task. Nevertheless, over the years many solutions were proposed [3–5, 14]. One natural source of ∞-categories, is categories enriched in topological spaces, where composition is strictly associative which trivially provides all "higher coherences" (namely, constant maps). Every ∞-category is in fact equivalent to one arising from a topologically enriched one. However, since all constructions in ∞-category theory need to be defined only up to coherent homotopy anyway, the language of topologically enriched categories is too rigid and hence inconvenient for developing an actual theory of ∞-categories. Thus, we shall adopt a rather indirect but very effective approach, using the notion of a *quasi-category*. This approach was initiated by A. Joyal [9, 10] and further developed by J. Lurie. Most of the definitions and basic results are from Lurie's seminal work [11].

4.2.2 Quasi-Categories

Recall that an ordinary category C can be encoded via a certain simplicial set $N\,(C) \in \mathrm{Set}^{\Delta^{op}}$ called the *nerve* of C. The vertices of $N\,(C)$ are the objects of C, the edges are the morphisms and the higher n-simplices encode composition in the following way. For every chain of composable morphisms in C:

$$X_0 \xrightarrow{f_0} X_1 \xrightarrow{f_1} \ldots \xrightarrow{f_{n-1}} X_n$$

we have a unique n-simplex σ such that $\sigma_{[i,i+1]} = f_i$. More concretely, if $h = g \circ f$ we have a unique 2-simplex

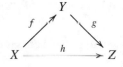

and the higher simplices are just uniquely inserted whenever their 2-skeleton is present. The face maps are obtained via composition in the obvious way, while the degeneracies insert identity morphisms. The simplicial identities correspond to the associativity and unitality axioms. The nerve construction constitutes a fully faithful embedding $\text{Cat} \hookrightarrow \text{Set}^{\Delta^{op}}$, so one might say that categories are just a particular kind of simplicial sets.

A quasi-category would be a simplicial set C which behaves somewhat like the nerve of a category, but where the conditions on the higher simplices that control composition are relaxed. We think of a 2-simplex in a quasi-category C

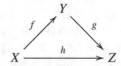

as exhibiting a *particular way* to compose g and f up to homotopy to obtain h. We shall require that whenever we have a simplicial subset of C of the form $X \xrightarrow{f} Y \xrightarrow{g} Z$, there shall always *exist* a 2-simplex in C as above, but unlike in the case of a nerve of a category, we do not require it to be *unique*. Now, a 3-simplex in C

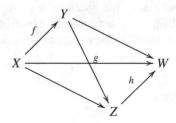

should be thought of as follows:

1. The vertices correspond to objects.
2. The edges correspond to morphisms.
3. The faces correspond to homotopies witnessing there particular choices of composition for adjacent morphisms.
4. The *interior* witnesses the associativity

$$h \circ (g \circ f) \sim (h \circ g) \circ f$$

(with respect to the particular choices of compositions given by the 2-dimensional faces)

Similarly the 4-simplices encode composable chains of 4-morphisms

$$X \xrightarrow{f} Y \xrightarrow{g} Z \xrightarrow{h} W \xrightarrow{k} U$$

with a choice of compositions up to specified homotopy, homotopies witnessing the associativity of the chosen compositions, and the *interior* encodes the Stasheff pentagon coherence for the chosen associativity homotopies. For $0 \le i \le n$ denote by $\Lambda_i^n \subset \Delta^n$ the simplical set obtained by removing the interior and the face opposing vertix i in Δ^n. Inspecting the ideas above systematically leads to the following definition:

Definition 1 A simplicial set $C \in \mathrm{Set}^{\Delta^{op}}$ is called a *quasi-category* if for every inner horn inclusion $\Lambda_i^n \hookrightarrow \Delta^n$ where $0 < i < n$ and a diagram

there exists a dashed arrow making the diagram commutative (this is known as the *inner horn filling condition*).   □

The first condition, for $n = 2$ and $i = 1$, means precisely that every pair of composable morphisms can indeed be composed up to homotopy. For $n = 3$ and $i = 1, 2$, the condition ensures that the compositions can be chosen to be associative up to homotopy. The higher conditions ensure that the associativity of the composition can be made *coherent* in a way which is itself coherent and so forth. We note that asking also for *uniqueness* of the extension in the diagram above for every $f : \Lambda_i^n \to C$ would be equivalent to asking that C is the nerve of an ordinary category. Thus, a quasi-category is strictly a generalization of an ordinary category.

We thus use categorical language when working with quasi-categories. Given a quasi-category $C \in \mathrm{Set}^{\Delta^{op}}$

1. We refer to the elements of the set C_0 as the *objects of C*.
2. We refer to the elements of the set C_1 as the *morphisms of C*.
3. Given a morphism $f \in C_1$ we call $d_1(f) \in C_0$ the *source of f* and call $d_0(f)$ the *target of f*. We write $f : d_1(f) \to d_0(f)$.
4. Given an object $c \in C_0$ we refer to the morphism $s_0(c) : c \to c$ as the *identity of c* and denote $\mathrm{Id}_c := s_0(c)$. Here $s_0 : C_0 \to C_1$ is the degeneracy map.
5. Given $f : a \to b$, $g : b \to c$ and $h : a \to c$ we write $g \circ f = h$ when there **exists** a 2-simplex $T \in C_2$ such that $d_2(T) = f$, $d_0(T) = g$ and $d_1(T) = h$.
6. The inner horn lifting condition implies that for every f and g as above there exists an h such that $g \circ f = h$. Note however, that for a general quasi-category C there can be two different edges $h \ne h'$ such that $g \circ f = h$ and $g \circ f = h'$.
7. In particular it is possible to have two different edges $h, h' : a \to b$ such that $h \circ \mathrm{Id}_a = h'$. In this case we say that *h and h' are homotopic* and write $h \sim h'$ the inner horn lifting condition implies that $\sim$ is an equivalence relation.
8. Given 3 morphisms $g \circ f = h$ and another morphism h' the inner horn lifting condition implies that $g \circ f = h'$ if and only if $h \sim h'$.

9. Given a morphism $f: a \to b$ in C a *section* of f is a morphism $s: b \to a$ such that $f \circ s = \mathrm{Id}_b$ and a *retraction* of f is a morphism $r: b \to a$ such that $r \circ f = \mathrm{Id}_a$.

10. If g is a both a section and a retract of f we say that g is an inverse of f and write $g = f^{-1}$. If f admits an inverse we will say that f *is an isomorphism*.

11. If a and b are objects in C such that there exists an isomorphism $f: a \to b$ we say that a and b are isomorphic and write $a \cong b$.

12. We call a quasi-category an ∞-groupoid if every morphism $f \in C$ is an isomorphism. Note that if C happens to be the nerve of a category C then it is an ∞-groupoid if and only if C itself is a groupoid.

It can be shown that being an ∞-groupoid can be described entirely in terms of lifting properties

Theorem 2 ([10]) *A simplicial set $C \in Set^{\Delta^{op}}$ is an ∞-groupoid if and only if for every horn inclusion $\Lambda_i^n \hookrightarrow \Delta^n$ where $0 \leq i \leq n$ and a diagram*

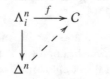

there exists a dashed arrow making the diagram commutative (this is known as the horn filling condition *or the* Kan condition*).* □

4.2.3 ∞-Groupoids and the Homotopy Hypothesis

One of the ways that the classical notion of a groupoid, i.e. a category in which all morphisms are invertible, is related to homotopy theory as follows. Given a space X, we can construct a groupoid $\Pi_{\leq 1} X$, known as the *fundamental groupoid* of X, by taking the points to be the objects, homotopy classes of paths to be the morphisms, and concatenation to be the composition rule. This construction manifestly depends only on the homotopy type of X. On the downside, it knows only about the connected components of X and their fundamental groups. In particular, for a simply connected space like S^2 we get the trivial groupoid. This can be fixed by constructing an ∞-*groupoid* in which instead of taking paths *up to* homotopy, we *remember* the homotopies.

Definition 3 The fundamental ∞-groupoid of a space X is the simplicial set $\mathrm{Sing}(X)$ given by $\mathrm{Sing}(X)_n = \hom_{\mathrm{Top}}(\Delta^n, X)$. Using Theorem 2 it is not hard to verify that $\mathrm{Sing}(X)$ is indeed an ∞-groupoid.

The objects and morphisms are again the points and paths, but a 2-simplex encodes how two paths can be concatenated and homotoped to a third path. □

Example 1 Let us consider for example the space S^2. In Sing (S^2) we will have for example an object $X \in$ Sing $(S^2)_0$ corresponding to a point on the equator, and a morphism $f: X \to X$ in Sing $(S^2)_1$ corresponding to the equator itself. The morphism f is homotopic to the identity of the object X (i.e. the constant path), but in at least two ways. One, is through the northern hemisphere, and the other, trough the southern hemisphere. Since those two homotopies are not homotopic to one another, their composite represents a non-contractible loop $S^1 \to$ Aut$_{\text{Sing}(S^2)}(X)$. This is precisely a generator of $\pi_2 S^2 \simeq \mathbb{Z}$.

A basic philosophical principle (which we make precise later) is that the fundamental ∞-groupoid contains in fact *all* the homotopical information of a homotopy type. The inverse process being the *geometric realization* functor. This is known as:

The Homotopy Hypothesis: The ∞-category of homotopy types is equivalent to the ∞-category of ∞-groupoids.

In particular every ∞-groupoid is equivalent to Sing(X) for some space X.

Remark 4 It follows that ∞-categories simultaneously generalize categories and homotopy types. Thus, they constitute a way to infuse homotopy theory in the very foundations of our categorical framework.   □

From this perspective, ∞-categories are categories enriched in ∞-groupoids, very much like ordinary categories are enriched in sets. It is now possible to say in which way a quasi-category C is indeed a model for an ∞-category. That is, a category coherently enriched in homotopy types. First, we need to construct the mapping spaces. The idea is to construct the simplicial set Map$_C (X, Y)$, whose n-simplices $\sigma: \Delta^n \to$ Map (X, Y) are precisely the $(n + 1)$-simplices $\tilde{\sigma}: \Delta^{n+1} \to C$, such that $\tilde{\sigma}(0) = X$ and $\tilde{\sigma}(1, \ldots n + 1)$ is the degenerate constant simplex on Y. The simplicial set Map$_C (X, Y)$ can be shown to be an ∞-groupoid and its homotopy type (or of its geometric realization) is the corresponding homotopy type of the space of morphisms from X to Y. With some more work one can also cook up canonical (up to homotopy) composition maps

$$\text{Map}(X, Y) \times \text{Map}(Y, Z) \to \text{Map}(X, Z),$$

homotopies witnessing associativity, higher homotopies witnessing coherence and so forth.

For a category C one can consider the subcategory $C^{\cong}$ consisting of all objects of C but only the isomorphisms. $C^{\cong}$ is then the maximal groupoid inside C. Similarly, given a quasi-category C one can take $C^{\cong}$ to be the sub simplicial set consisting of only simplices with all edges isomorphisms. It is can be checked that $C^{\cong}$ is an ∞-groupoid. The ∞-groupoid $C^{\cong}$ can be thought of as "the space of objects in C".

4.2.4 Quasi-Categories from Topological Categories

As mentioned above, topologically enriched categories should also provide examples of ∞-categories, and thus of quasi-categories. First, for every simplex Δ^n we construct a certain topologically enriched category $\mathfrak{C}[\Delta^n]$.

Definition 5 The objects of $\mathfrak{C}[\Delta^n]$ are the elements of $[n] := \{0, \ldots, n\}$. The morphisms are defined as follows:

- For $i > j$ we have $\text{Map}_{\mathfrak{C}[\Delta^n]}(i, j) = \varnothing$.
- $\text{Map}_{\mathfrak{C}[\Delta^n]}(i, i) = \{\text{Id}_i\}$.
- For $i < j$ we have $\text{Map}_{\mathfrak{C}[\Delta^n]}(i, j) = I^{\{i+1,\ldots,j-1\}}$, where $I = [0, 1]$ is the unit interval.   □

Composition is defined by the natural inclusion

$$I^{\{i+1,\ldots,j-1\}} \times I^{\{j+1,\ldots,k-1\}} \cong I^{\{i+1,\ldots,\hat{j},\ldots,k-1\}} \subseteq I^{\{i+1,\ldots,k-1\}}$$

where the last inclusion is the identification of $I^{\{i+1,\ldots,\hat{j},\ldots,k-1\}}$ with the subset of $I^{\{j+1,\ldots,k-1\}}$, of tuples with j-th entry equal to 0.

Example 2

- $\mathfrak{C}[\Delta^0] = *$ is the category associated with a singleton.
- $\mathfrak{C}[\Delta^1]$ has two object and a unique morphism between them.
- $\mathfrak{C}[\Delta^2]$ has 3 objects c_0, c_1, c_2, a pair of maps $f_{0,1} : c_0 \to c_1$, $f_{1,2} : c_1 \to c_2$ and a path from $f_{1,2} \circ f_{0,1}$ to a new morphism $f_{0,2}$.
- $\mathfrak{C}[\Delta^3]$ has 4 objects c_0, c_1, c_2, c_3, maps $f_{i,j}$ for $0 \le i < j \le 3$ with $f_{i,j} : c_i \to c_j$. Moreover, we have paths $h_{0,1,2}, h_{0,1,3}, h_{0,2,3}$ and $h_{1,2,3}$ with $h_{i,j,k} \in \text{Map}(c_i, c_k)$ the a path connecting $f_{j,k} \circ f_{i,j}$ to $f_{i,k}$. Finally, we have a square $h_{0,1,2,3}$ of morphisms which serve as a homotopy between the paths of morphisms in $\text{Map}_{\mathfrak{C}[\Delta^3]}(c_0, c_3)$ that looks like this:

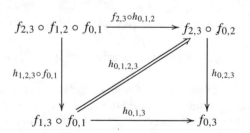

- In general, the space of maps $\text{Map}_{\mathfrak{C}[\Delta^n]}(c_i, c_j)$ will be a topological $j - i - 1$-dimensional cube, and the composition will embed the product of smaller cubes as a face of a larger cube.

The *coherent nerve* of a topological category is defined by $N^\Delta(C)_n = \hom(\mathfrak{C}[\Delta^n], C)$ and is always a quasi-category. This in particular allows us to define the ∞-category of homotopy types:

Definition 6 Let CW be the topological category of CW-complexes with mapping spaces given the compact-open topology. The ∞-category S of homotopy types is the coherent nerve of CW. □

Remark 7 We note that ∞-categories themselves can be organized in an ∞-category Cat_∞ and if we restrict to the full subcategory of ∞-groupoids Grpd_∞, we have a manifestation of the homotopy hypothesis as the equivalence of ∞-categories $S \simeq \mathrm{Grpd}_\infty$. □

4.2.5 ∞-*Category Theory*

With the notion of an ∞-category formalized precisely by the notion of a quasi-category, we can now do some actual "∞-category theory". One of the advantages of quasi-categories over other models is that a functor of quasi-categories is simply a map of simplicial sets. The data of such a map already encodes preservation of composition up to coherent homotopy. Better yet, the category of simplicial sets is Cartesian closed. The product of two quasi-categories $C \times \mathcal{D}$ is a quasi category itself, which should be thought of as the product of the ∞-categories. Similarly, the internal hom of simplicial sets $\underline{\hom}(C, \mathcal{D})$ is itself a quasi-category [11, Proposition 1.2.7.3.] (this is far less trivial fact), and should be thought of as the ∞-category of functors from C to $\mathcal{D}$. We thus denote $\underline{\hom}(C, \mathcal{D})$ by $\mathrm{Fun}(C, \mathcal{D})$. There is also an obvious notion of equivalence for ∞-categories. That is a functor $F: C \to \mathcal{D}$, such that there exists a functor $G: \mathcal{D} \to C$ and homotopies

$$GF \sim \mathrm{Id}_C \in \mathrm{Fun}(C, C), \qquad FG \sim \mathrm{Id}_{\mathcal{D}}, \in \mathrm{Fun}(\mathcal{D}, \mathcal{D}).$$

In fact, as mentioned above, there is a quasi-category Cat_∞ whose objects are quasi-categories and edges are functors.

The next step is to consider *(co)limits*. The most basic (co)limits are initial and terminal objects. To define them, we first define *over* and *under* categories as follows. For an object $X \in C$, we define $C_{X/}$ to a simplicial set for which the n-simplices $\sigma: \Delta^n \to C_{X/}$ are in bijection with $\tilde{\sigma}: \Delta^{n+1} \to C$ such that $\tilde{\sigma}(0) = X$. Note that there is a canonical projection functor $p: C_{X/} \to C$ given by restriction to the face opposite to the 0-th vertex. The objects of $C_{X/}$ are precisely maps $X \to Y$ and the morphisms are maps under X up to given homotopy. Even better, the fiber of p over Y is precisely the space (∞-groupoid) of morphism from X to Y as defined before. We define X to be *initial* if p is an equivalence. Similarly we can define the over category $C_{/X} \to C$ and a terminal object. The full ∞-subcategory spanned by initial (resp. terminal) objects is a ∞-groupoid (i.e. a "homotopy type"). It is a theorem that this homotopy type is either empty or contractible. It is a general principle in ∞-categories, that uniqueness is replaced by a "contractible space of choices". Now, we define colimits (resp. limits) of general diagrams $F: K \to C$ as initial (resp. terminal) objects in the corresponding ∞-categories of cones $C^{F/}$ (resp. $C^{/F}$). One way to think about a cone on a diagram $F: K \to C$ is as an natural transformation to a constant functor. This leads to a the definition of $C^{F/}$ as the pullback

$$
\begin{array}{ccc}
C^{F/} & \longrightarrow & \text{Fun}\left(K \times \Delta^1, C\right) \\
\downarrow & & \downarrow \\
C & \xrightarrow{c_0 \mapsto (F, \text{const}(c_0))} & \text{Fun}\left(K, C\right) \times \text{Fun}\left(K, C\right)
\end{array}
$$

Where for $c_0 \in C$ the functor $\text{const}(c_0) \in \text{Fun}\left(K, C\right)$ is the constant functor with value c_0.

Remark 8 The (co)limits of functors of ∞-categories correspond to what is classically known as *homotopy (co)limits*, but instead of defining them as derived functors of some strict 1-categorical constructions, we define them intrinsically using a coherent version of the corresponding universal property.   □

Similarly to the theory of ordinary categories, one can also define *adjoint functors* and show they respect (co)limits. In particular, we have the functor

$$\text{const}: \mathcal{D} \to \text{Fun}\left(C, \mathcal{D}\right)$$

taking every object of $\mathcal{D}$ to the constant C-shaped diagram on this object. If $\mathcal{D}$ admits all C-shaped colimits (resp. limits), its left (resp. right) adjoint will be precisely the operation of taking the colimit (resp. limit) of the diagram. Finally, we note that we also have an ∞-categorical analogue of the *Yoneda lemma*. Since we now have mapping spaces instead of mapping sets, our representable functors have target S and not Set. The Yoneda embedding takes the form of a functor of ∞-categories

$$\text{よ}: C \to \text{Fun}\left(C^{op}, S\right)$$

and it is fully faithful in the sense that it induces homotopy equivalences on mapping spaces.

4.2.6   The Homotopy Category

As we mentioned above the collection of all ∞-categories can be given a structure of an ∞-category Cat_∞. The objects of Cat_∞ are themselves (small) ∞-categories and for $C, \mathcal{D} \in \mathrm{Cat}_\infty$ we have an equivalence of homotopy types:

$$\mathrm{Map}_{\mathrm{Cat}_\infty}(C, \mathcal{D}) \sim \mathrm{Fun}(C, \mathcal{D})^{\cong}.$$

The nerve construction can be then considered as fully-faithful embedding

$$\mathrm{Cat}_1 \hookrightarrow \mathrm{Cat}_\infty$$

of the $(2, 1)$-category of 1-categories to the ∞-category of ∞-categories. This embedding admits a left adjoint

$$h : \mathrm{Cat}_\infty \to \mathrm{Cat}_1$$

$$C \mapsto hC.$$

The ordinary category hC is called the *homotopy category of C* and can be described as follows. hC has the same objects as hC and we have

$$\mathrm{Hom}_{hC}(X, Y) = \pi_0(\mathrm{Map}_C(X, Y)).$$

Note that in particular $h\mathcal{S}$ is equivalent to the homotopy category of CW-complexes. Also for a space X we have an equivalence of groupoids

$$h\mathrm{Sing}(X) = \Pi_{\leq 1}(X)$$

4.2.7   ∞-Categories and Homological Algebra

The language of ∞-categories turn out to naturally encompass homological algebra. This relationship gives an important source for ∞-categories as well as greatly simplifies and conceptualises classical results from homological algebra. In particular:

Definition 9 (see [12] 1.3.1.6) Let R be a ring, we define the ∞-category $D_\infty(R)$ as follows: For each $n \geq 0$ let the set of n-simplices $D_\infty(R)_n$ be the set of all ordered pairs

$$(\{X_i\}_{0 \leq i \leq n}, \{f_I\}_{I \subset [n], 2 \leq |I|})$$

where for each $0 \leq i \leq n$, X_i is a bounded from below complex of projective R-modules and for each $I = \{i_- < i_1 < i_2 < \cdots < i_m < i_+\}$ the map f_I is a degree m-map

$$f_I : X_{i_-} \to X_{i_+}$$

satisfying the equation

$$dx_{i_+} \circ f_I - (-1)^m f_I \circ dx_{i_-} = \sum_{1 \leq j \leq m} (-1)^j (f_{I-i_j} - f_{\{i_j < \cdots < i_m < i_+\}} \circ f_{\{i_- < i_1 < \cdots < i_j\}}).$$

Note that the f_I are not assumed to be chain maps in general.

Lurie proved that $D_\infty(R)$ is a quasi-category [12, 1.3.1.10] and that $hD_\infty(R)$ is the usual (bounded from below) derived category of R. In particular, unwinding the definitions gives

1. Vertices of $D_\infty(R)$ are complexes of projective R-modules, bounded from below.
2. 1-simplices of $D_\infty(R)$ consist of two such complexes X, Y and a chain map $f : X \to Y$.
3. 2-simplices of $D_\infty(R)$ consist of three such complexes X, Y, Z, a triple of chain maps $f : X \to Y, g : Y \to Z, h : X \to Z$ and a chain homotopy between $g \circ f$ to h.

Given any ∞-category C which has all colimits and an object $c_0 \in C$ we define a functor

$$c_0 \otimes - : S \to C$$

$$X \mapsto \operatorname*{colim}_{X} \operatorname{const}(c_0)$$

where the colimit is taken for the functor $\operatorname{const}(c_0) : X \to C$ constant on c_0. Consider the object $R \in D_\infty(R)$ where R is taken to be the complex with R in degree 0 and 0 everywhere else.

We get a functor

$$R \otimes - : S \to D_\infty(R).$$

This is a reformulation of a familiar construction. The functor $R \otimes -$ is equivalent to the one sending a CW-complex X to its singular chains $C_*(X, R)$. In particular the Mayer-Vietoris theorem can now be restated as saying that the ∞-functor $R \otimes -$ preserves pushouts.

4.2.8 Stable ∞-Categories

The category $hD_\infty(R)$ is known classically to admit a special structure of a **triangulated category**. In particular all hom-sets in $hD_\infty(R)$ carry a structure of an abelian group and one can make sense of "exact triangles". Lurie showed [12] that all of this **structure** on $hD_\infty(R)$ comes naturally from a **property** of $D_\infty(R)$.

Definition 10 (see [12]) We say that an ∞-category C is *stable* if

1. C admits all finite limits and colimits.
2. The initial and terminal objects in C are isomorphic.
3. Every commutative square $S \in \mathrm{Fun}(\Delta^1 \times \Delta^1, C)$ is a pushout square if and only if it is a pullback square.   □

For every ring R the ∞-category $D_\infty(R)$ is stable and Lurie showed [12] in general that if $C \in \mathrm{Cat}_\infty$ is stable then hC is naturally triangulated.

Stable ∞-categories can be considered as the ∞-analogues of abelian categories and have the advantage of having a "linear" nature. The colimit preserving functor of singular chains

$$\mathbb{Z} \otimes -: \mathcal{S} \to D_\infty(\mathbb{Z}),$$

can thus be considered as a way to "linearize" the homotopical information in a space X. In fact there exists a universal example of a stable ∞-category Sp together with a colimit preserving functor

$$\Sigma_+^\infty : \mathcal{S} \to \mathrm{Sp}.$$

The stable ∞-category Sp is known as the ∞-category of spectra and it has great importance in modern homotopy theory. In particular let $\mathbb{S}^n$ denote the pullback

$$
\begin{array}{ccc}
\mathbb{S}^n & \longrightarrow & \Sigma_+^\infty \emptyset, \\
\downarrow & & \downarrow \\
\Sigma_+^\infty S^n & \longrightarrow & \Sigma_+^\infty *
\end{array}
$$

we get that the abelian group $\mathrm{Hom}_{h\mathrm{Sp}}(\mathbb{S}^n, \Sigma_+^\infty X)$ is the n'th stable homotopy group of X.

4.2.9   Localization

One useful way to create new ∞-categories from old ones is by inverting a collection of morphisms, a process known as *localization*. More precisely we have the following theorem:

Theorem 11 *Let C be a quasi-category and $\mathcal{W} \subset C_1$ be a collection of morphisms. Then there exist an ∞-category $C[\mathcal{W}^{-1}]$ and map*

$$C \to C[\mathcal{W}^{-1}]$$

such that for every ∞-category $\mathcal{D}$ the induced map

$$\mathrm{Fun}(C[W^{-1}], \mathcal{D}) \to \mathrm{Fun}(C, \mathcal{D})$$

is fully-faithfull and a functor

$$G \in \mathrm{Fun}(C, \mathcal{D})$$

is in the essential image if and only if for every $e \in W$ the morphism $G(e) \in \mathcal{D}_1$ is an isomorphism.   □

An important aspect of this construction is that even when C happens to be an ordinary category $C[W^{-1}]$ is very often not. In fact Barwick and Kan proved [3] that **every** ∞-category is of the form $C[W^{-1}]$ for some category C and a subset of morphisms $W \subset C_1$.

In particular,

1. Let C be a category then $C[C_1^{-1}]$ is an ∞-groupoid which is equivalent to $\mathrm{Sing}(|N(C)|)$.
2. Let Top be the category of topological spaces and $W \subset \mathrm{Top}_1$ the collection of weak-homotopy equivalences then $\mathrm{Top}[W^{-1}]$ is equivalent to $\mathcal{S} \simeq \mathrm{Grpd}_\infty$.
3. Let QCat $\subset \mathrm{Set}^{\Delta^{op}}$ be the full subcategory of $\mathrm{Set}^{\Delta^{op}}$ spanned by on quasi-categories and let $W \subset \mathrm{Qcat}$ be the collection of categorical equivalences then $\mathrm{QCat}[W^{-1}]$ is equivalent to Cat_∞.
4. For a ring R let $\mathrm{Ch}(R)$ be the category of bounded below complexes of R-modules. Let $W \subset \mathrm{Ch}(R)$ be the collection of quasi-isomorphisms, then $\mathrm{Ch}(R)[W^{-1}]$ is equivalent to $D_\infty(R)$.

Given a category C and a collection of morphisms $W \subset C$ it is in general quite hard to describe the ∞-category $C[W^{-1}]$. However it can be easier in the presence of additional assumptions and structure. One such setup is that of a *model category* which adds to $W \subset C$ two additional collections of morphisms $\mathcal{F}\mathrm{ib}, \mathrm{Cof} \subset C$ called fibrations and cofibrations such that $\mathcal{F}\mathrm{ib}, \mathrm{Cof}, W$, and C satisfy certain axioms. Using model categories is a very useful and ubiquitous way to describe and work in ∞-categories but we shall not discuss them further in these notes.

4.3 ∞-Topoi

4.3.1   Definitions

We shall be interested in ∞-topoi, which are a particular kind of ∞-categories which behave like ∞-categories of sheaves on a *topological space*, taking values in homotopy types (i.e. objects of the ∞-category $\mathcal{S}$). There is an "intrinsic" characterization of ∞-topoi given by a set of axioms on the ∞-category, but instead we shall give an "extrinsic" one by explaining where ∞-topoi come from. An archetypal example of an ∞-topos is S itself, which is "sheaves on a point". Another example is any presheaves (again of homotopy types) ∞-category

$$\text{Psh}\,(C) = \text{Fun}\,(C^{op}, S)$$

where C is some (small) ∞-category. This is just the ∞-category of C^{op}-shaped diagrams of spaces. Given a topological space X, we can construct the category Op_X whose objects are the open subsets of X and morphisms are inclusions. The ∞-category Sh (X) of sheaves of homotopy types on X, is the full ∞-subcategory of the presheaf ∞-category

$$\text{Psh}\,(X) = \text{Fun}\,(Op_X^{op}, S)$$

spanned by objects which satisfy a certain ∞-categorical analogue of the sheaf condition. Since the category of sets is fully faithfully embedded in S as the homotopically discrete homotopy types, we also have a full ∞-subcategory of Sh (X) spanned by those functors having essential image in Set $\subseteq S$. This full subcategory is the classical category of sheaves of sets on X. This construction works more generally for constructing sheaves on a site (i.e. a category with a *Grothendieck topology*).

The inclusion Sh $(X) \hookrightarrow$ Psh (X) admits a left adjoint, which is left exact (i.e. preserves finite limits). It turns out that being an ∞-topos is *equivalent* to being a full subcategory of some presheaf category, such that the inclusion admits a left exact left adjoint. In other words, we can take the following as a definition:

Definition 12 An ∞-*topos* is an ∞-category that is a left exact reflective localization of a presheaf ∞-category. $\square$

There is a somewhat subtle but crucial point to note. Given a topological space X, we can construct from it an ∞-groupoid Sing (X), which is in particular an ∞-category, and consider presheaves on that

$$\text{Psh}\,(\text{Sing}\,(X)) = \text{Fun}\,(\text{Sing}\,(X)^{op}, S)\,.$$

This is an ∞-topos and is another notion of "sheaves of homotopy types on X", but it is very different from the ∞-topos Sh (X) that we have constructed above. For example, if $x, y \in X$ are two points connected by a path, then they are isomorphic objects in Sing (X) and thus, every presheaf on Sing (X) must assign them the same value. In construct, a sheaf on X may have different stalks at different points of the same path connected component. In particular, one may have very interesting sheaves on the contractible space $\mathbb{R}^n$. When we restrict attention to sheaves of *sets*, we see that Psh $(\text{Sing}\,(X), \text{Set})$ is equivalent to Fun $((\Pi_{\leq 1} X)^{op}, \text{Set})$, which by covering space theory corresponds to locally-constant sheaves. Thus, Psh $(\text{Sing}\,(X))$ may be thought of as the locally-constant "up to coherent homotopy" sheaves of homotopy types on X.

Now that we have defined ∞-topoi, we would like to consider morphisms between them. A map of topological spaces $f : X \to Y$ induces a pullback functor

$$f^* : \mathrm{Sh}\,(Y) \to \mathrm{Sh}\,(X)\,.$$

This functor is both left exact and admits a right adjoint

$$f_* : \mathrm{Sh}\,(X) \to \mathrm{Sh}\,(Y)\,.$$

We adopt this a definition of a morphism of ∞-topoi:

Definition 13 A morphism of topoi $f : \mathcal{X} \to \mathcal{Y}$ (called *geometric morphism*) is an adjunction $f^* : \mathcal{Y} \leftrightarrows \mathcal{X} : f_*$ such that f^* is left exact. □

Furthermore, ∞-topoi and geometric morphisms can be organized into an ∞-category Topoi_∞. The terminal object is $\mathcal{S}$ (which we indeed think of as corresponding to the "one point space"). The right adjoint $c_* : \mathcal{X} \to \mathcal{S}$ functor in the unique geometric morphism $c : \mathcal{X} \to \mathcal{S}$ should be thought of as taking *global sections* and its left adjoint $c^* : \mathcal{S} \to \mathcal{X}$ as the *constant sheaf* functor

4.3.2   The Shape of an ∞-Topos

The construction $A \mapsto \mathrm{Psh}\,(A)$ extends to a fully faithful embedding $\mathcal{S} \hookrightarrow \mathrm{Topoi}_\infty$, so we might identify homotopy types with a special kind of ∞-topoi. Not every ∞-topos is of this form, but we can try to *approximate* it by such. This process should generalize (or at least be analogous to) taking a nice topological space (like a CW-complex) and constructing its homotopy type. To construct such an approximation, we can start with a simpler task of recovering the homotopy type A from the ∞-category $\mathrm{Psh}\,(A)$. Again, we take our cue from the discrete case. When A is a *set* viewed as a discrete homotopy type, we can recover A from $\mathrm{Fun}\,(A^{op}, \mathcal{S})$ by taking the colimit of the constant presheaf on a point pt_A. Namely,

$$\operatorname*{colim}_A \left(\mathrm{pt}_A\right) = \coprod_{a \in A} \mathrm{pt} \simeq A$$

This in fact works for *any* homotopy type A. That is, every homotopy type is the (homotopy) colimit of the constant A-shaped diagram on a point. We can rephrase this slightly by using the fact that taking colimit is the adjoint functor of the constant diagram functor. Let

$$c_! : \mathrm{Fun}\,\left(A^{op}, \mathcal{S}\right) \to \mathcal{S}$$

be the left adjoint of the canonical functor

$$c^* : S \to \mathrm{Fun}\left(A^{op}, S\right).$$

What we get is

$$A \simeq A^{op} \simeq c_!\left(\mathrm{pt}_A\right)$$

We can now try to use the same idea to produce a homotopy type from a general ∞-topos X. We immediately run into the problem that $c_!$ is not part of the geometric morphism $c : X \to S$, but rather a consequence of the very special fact that c^* happened to admit a left adjoint. This is usually not the case for a general ∞-topos as a necessary (and by the adjoint functor theorem, sufficient) condition for c^* to admit a left adjoint, is that c^* preserves all limits, and this generally fails. It is true though that c^* preserves *finite limits*. This can be used to produce a *pro-object* in S, a notion that we now explain. When the ∞-category C has all small limits, the essential image of the Yoneda embedding

$$C^{op} \hookrightarrow \mathrm{Fun}\,(C, S)$$

consists only of functors that preserve all limits. The Yoneda embedding thus in particular factors through the full subcategory of *finite-limit* preserving functors

$$C \hookrightarrow \mathrm{Pro}\,(C) = \mathrm{Fun}^{\mathrm{fin}}\,(C, S)^{op} \subseteq \mathrm{Fun}\,(C, S)^{op}.$$

The ∞-category $\mathrm{Pro}\,(C)$ can be thought of as a "mild" extension of C. This ∞-category has an alternative description in terms of formal filtered limits of objects in C. For example, when C is the ordinary category of *finite groups*, $\mathrm{Pro}\,(C)$ is equivalent to the ordinary category of *profinite groups*.

Back to our business, the fact that c^* preserves finite limits and some abstract ∞-categorical facts about adjoints allow us to define a "pro left adjoint"

$$c_! : \mathrm{Fun}^{\mathrm{fin}}\,(X, S) \to \mathrm{Fun}^{\mathrm{fin}}\,(S, S) \simeq \mathrm{Pro}\,(S)^{op}$$

as the pre-composition with c^*.

Definition 14 Let X be an ∞-topos. The pro-homotopy type $c_!\left(\mathrm{pt}_X\right) \in \mathrm{Pro}\,(S)$ is called the *shape* of X, which we denote by $|X|$. $\square$

In fact a simple manoeuvre with adjunctions shows that

$$|X| = c_* \circ c^* : S \to S.$$

As we promised in the introduction, there is also a *relative* version of this story. Let $t: X \to B$ be a geometric morphism of ∞-topoi where B is some "base" ∞-topos, which need not be S. The left adjoint $t^*: B \to X$ has a "pro left adjoint"

$$t_! : \mathrm{Pro}\,(X) \to \mathrm{Pro}\,(B)\,.$$

The *relative shape* of X (or the shape of X relative to B) is defined to be the image of the terminal object of $\mathrm{Pro}\,(X)$ under $t_!$ and denoted by $|X/B|$. Similar manoeuvres to the above give

$$|X/B| = c_* \circ t^*: B \to S,$$

where $c_*: X \to S$ is the global sections fucntor.

Now given a commutative triangle

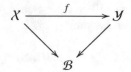

of ∞-topoi, we obtain a map $\overline{f}: |X/B| \to |Y/B|$. We shall be mostly interested in the special case $Y = B$ and in finding sections to f. In this case $|X/Y|$ is some object of $\mathrm{Pro}\,(Y)$ and $|Y/Y|$ is the terminal object $\mathrm{pt} \in \mathrm{Pro}(Y)$. Thus, if we think of $|X/Y|$ as a certain sheaf of pro-homotopy types on Y, then a section of $\overline{f}$ corresponds to a *global section* of this sheaf.

4.4 Obstruction Theory

Recall that our main goal was to establish an algebraic obstruction theory for the existence of sections to a certain map of "spaces" $f: X \to Y$. When f is a geometric morphism of ∞-topoi, we can use the theory of relative shape, to construct a map $\overline{f}: |X/Y| \to |Y/Y|$ of sheaves of pro-objects in Y to which we can apply the machinery of *Postnikov truncations*. Recall that we think of Y as the ∞-category of sheaves of homotopy types on a topological space. Thus, $\mathrm{Pro}\,(Y)$ is like sheaves of pro-homotopy types on some topological space. For simplicity, we begin with the obstruction theory for maps of homotopy types and then indicate how the theory generalizes to $\mathrm{Pro}\,(Y)$.

4.4.1 Obstruction Theory for Homotopy Types

Let $f: Y \to X$ be a map of homotopy types. Denote the homotopy fiber of f by F and assume for simplicity that X and F are connected. Note that $\pi_1 X$ acts naturally on the homotopy groups of F and we can therefore consider the corresponding cohomology groups $H^t(X, \pi_s F)$ in the sense of cohomology with local coefficients. Our immediate goal is to construct a sequence of elements

$$o_n \in H^{n+1}(X, \pi_n F)$$

such that the vanishing of them would be equivalent to the existence of section (up to homotopy) of f. More precisely, the obstructions form a "tree" and the existence of a section will correspond to an infinite branch in this tree. Let us explain more precisely what exactly do we mean by that. The first obstruction

$$o_1 \in H^2(X, \pi_1 F)$$

is canonically defined and if $o_1 \neq 0$, then there is no section. If $o_1 = 0$, it means that it is a coboundary, and given a **choice** of a 1-cochain α_1 with $d\alpha_1 = o_1$, we will be able to produce a new class

$$o_2 \in H^3(X, \pi_2 F).$$

If $o_2 \neq 0$, then we are stuck and we can not produce a section using α_1, but we might be able to if we replace α_1 with some other 1-cochain with the same coboundary. Namely, o_2 is defined only if $o_1 = 0$ and depends on **how** it is zero. If $o_2 = 0$, we can again choose a 2-cochain exhibiting o_2 as a coboundary, and extract from it a new class

$$o_3 \in H^4(X, \pi_3 F).$$

If we are able to construct a compatible sequence of choices of coboundaries exhibiting the vanishing of all o_n-s, then we shall be able to construct a section. Conversely, if every such sequence terminates with a non-vanishing obstruction at some finite stage, then no section exists.

Remark 15 In fact, more can be said. One can consider the *space* of sections sec (f) and if this space is non-empty, there is a spectral sequence of the form

$$E_2^{s,t} = H^s(X, \pi_t F) \implies \pi_{t-s}(\sec(f)).$$

Note that the question of whether sec $(f) = \varnothing$ is in some sense a question about "$\pi_{-1}(\sec(f))$" which fits nicely with where the obstructions o_n live.   □

A fundamental observation we are about to use throughout is the so-called *Grothendieck construction* which in its most basic form, states that for every $X \in \mathcal{S}$, there is a canonical equivalence of ∞-categories:

$$\mathcal{S}_{/X} \simeq \operatorname{Fun}(X, \mathcal{S})$$

that goes as follows. Given a map $f : Y \rightarrow X$ of homotopy types, we can construct a functor of ∞-categories $\Phi : X \rightarrow \mathcal{S}$ (where X is viewed as an ∞-groupoid) by assigning to each point $x \in X$, the homotopy fiber $f^{-1}(x) \in \mathcal{S}$ and to each path $\gamma : x \rightarrow y$, the "parallel transport" $\gamma_* : f^{-1}(x) \rightarrow f^{-1}(y)$, and so on. In the other direction, given a functor $\Phi : X \rightarrow \mathcal{S}$, we can take its colimit $\operatorname*{colim}_{X}(F)$, which maps canonically to $\operatorname*{colim}_{X}(\mathrm{pt}) \simeq X$. These constructions are mutually inverse up to homotopy.

Example 3 When considering only the case where the fiber F is discrete, we obtain an equivalence between the category of covering spaces of X and the category of functors $\Pi_{\leq 1} X \rightarrow \operatorname{Set}$ or equivalently $\pi_1(X)$-Sets. This is a familiar fact from covering space theory. The above construction generalises this from discrete F to arbitrary F, which requires replacing the category Set with the ∞-category $\mathcal{S}$ and the groupoid $\Pi_{\leq 1} X$ with the ∞-groupoid X itself.

Given a map of homotopy types $f : Y \rightarrow X$, we assume for simplicity that X is connected and thus all homotopy fibers of f are homotopy equivalent to some space F. Thus, we can think of the corresponding functor $\Phi_f : X \rightarrow \mathcal{S}$ as assuming the value F at each point, but is generally non-constant, as loops in X can induce non-trivial self homotopy equivalences of F. One example is the identity map $\operatorname{Id} : X \rightarrow X$. Since all fibers of Id are contractible, Φ_{Id} is equivalent to the constant functor $\mathrm{pt}_X : X \rightarrow \mathcal{S}$. Thus, a section of a map $f : Y \rightarrow X$ corresponds to a natural transformation $\mathrm{pt}_X \Rightarrow \Phi_f$. This is quite literally a "compatible choice of a point in each fiber". To construct such a section we can adopt an inductive procedure which filters F by homotopically truncated pieces.

Postnikov Truncation: For every $n \geq 0$, we have functors $P_n : \mathcal{S} \rightarrow \mathcal{S}$ such that

$$\pi_k(P_n F) = \begin{cases} \pi_k(F) & k \leq n \\ 0 & k > n \end{cases}$$

Moreover, there are natural transformations $\operatorname{Id} \rightarrow P_n$, such that $F \rightarrow P_n F$ induces isomorphisms on π_k for $k \leq n$. Consequently, we obtain a tower decomposition $F \xrightarrow{\sim} \lim_n P_n F$.

We note that for $n \geq 1$ the homotopy fiber of $P_n F \rightarrow P_{n-1} F$ is the Eilenberg-MacLane space $K(\pi_n F, n)$, which is concentrated in a single homotopy degree n and as such, determined by the single group $\pi_n F$. Since the P_n-s are functors, we can apply this truncation procedure also to families of spaces parameterized by X

such as $\Phi \colon X \to S$. In particular, using the tower

$$\Phi \simeq \lim \left(P_n \circ \Phi \right)$$

we have a tower decomposition of the space of sections

$$\mathrm{Nat} \left(\mathrm{pt}_X, \, \Phi \right) \simeq \lim_n \mathrm{Nat} \left(\mathrm{pt}_X, \, P_n \circ \Phi \right).$$

If we are only interested in the *existence* of a section, we can try to inductively lift a point $\alpha_{n-1} \in \mathrm{Nat} \left(\mathrm{pt}_X, P_{n-1} \circ \Phi \right)$ to a point $\alpha_n \in \mathrm{Nat} \left(\mathrm{pt}_X, P_n \circ \Phi \right)$. For this, consider the pullback diagram (of functors $X \to S$)

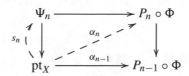

By the universal property of the pullback, a lift α_n is the same as a section s_n. The inductive step is thus exactly of the same form as our original problem, only that the functor $\Phi \colon X \to S$ is now replaced by the functor $\Psi_n \colon X \to S$. At each point $x \in X$, the value of Ψ_n is equivalent to the fiber of the map $P_n \left(F \right) \to P_{n-1} \left(F \right)$, which is the Eilenberg-MacLane space $K \left(\pi_n \left(F \right), n \right)$. This is a very simple space as it is concentrated entirely in homotopical degree n and as such it is completely determined by the single group $\pi_n \left(F \right)$ and the integer n. In other words, the functor Ψ_n factors through the connected component of $K \left(\pi_n \left(F \right), n \right) \in S$. As with ordinary groupoids, this connected component is equivalent to the classifying space the topological group of self equivalences $\mathrm{Aut} \left(K \left(\pi_n \left(F \right), n \right) \right)$. The functor Ψ_n is hence just a map of homotopy types

$$\Psi_n \colon X \to B\mathrm{Aut} \left(K \left(\pi_n \left(F \right), n \right) \right).$$

Now, giving a section s_n is the same as choosing a point in $\Psi_n(x)$ for every $x \in X$ in a compatible way. Thus, it can be shown to be the same as lifting the functor $\Psi_n \colon X \to S$ to a fucntor $\Psi_n \colon X \to S_* = S_{\mathrm{pt}/}$ where S_* is the ∞-category of pointed spaces. In maximal sub-∞-groupoid $S_*^{\simeq}$, the connected component that corresponds to the pointed space $K \left(\pi_n \left(F \right), n \right)$ is precisely $B\mathrm{Aut}_* \left(K \left(\pi_n \left(F \right), n \right) \right)$, the space of *base point preserving* self-equivalences of $K \left(\pi_n \left(F \right), n \right)$. Thus a section s_n is precisely a lift

$$B\mathrm{Aut}_* \left(K \left(\pi_n \left(F \right), n \right) \right)$$

$$X \xleftarrow{\hspace{2cm}} \xrightarrow{\Psi_n} B\mathrm{Aut} \left(K \left(\pi_n \left(F \right), n \right) \right),$$

Note that evaluation at a base point gives a fiber sequence

$$\text{Aut}_* \left(K \left(\pi_n \left(F \right), n \right) \right) \xrightarrow{\Omega g} \text{Aut} \left(K \left(\pi_n \left(F \right), n \right) \right) \to K \left(\pi_n \left(F \right), n \right).$$

By standard algebraic topology, the space $\text{Aut}_* \left(K \left(\pi_n \left(F \right), n \right) \right)$ is actually *discrete* and equivalent to $\text{Aut}_{\text{Grp}} \left(\pi_n \left(F \right) \right)$. Since the first map is a map of *groups*, we can apply the classifying space functor $B \left(- \right)$ and get the following fiber sequence of spaces

$$K \left(\pi_n \left(F \right), n \right) \to B\text{Aut}_{\text{Grp}} \left(\pi_n \left(F \right) \right) \xrightarrow{g} B\text{Aut} \left(K \left(\pi_n \left(F \right), n \right) \right).$$

Using the long exact sequence in homotopy groups we get:[1]

$$\pi_k \left(B\text{Aut} \left(K \left(\pi_n \left(F \right), n \right) \right) \right) = \begin{cases} \text{Aut}_{\text{Grp}} \left(\pi_n \left(F \right) \right) & k = 1 \\ \pi_n \left(F \right) & k = n + 1 \\ 0 & else \end{cases}$$

Namely, $B\text{Aut} \left(K \left(\pi_n \left(F \right), n \right) \right)$ is almost the same as the Eilenberg-MacLane space $K \left(\pi_n \left(F \right), n + 1 \right)$ except that it has a fundamental group $\text{Aut}_{\text{Grp}} \left(\pi_n \left(F \right) \right)$. This fundamental group acts naturally on the π_{n+1} of the space, which is $\pi_n \left(F \right)$. Just like ordinary Eilenberg-MacLane spaces represent cohomology with *constant* coefficients, this space represents cohomology with *twisted* coefficients. To see this, note that the map s from $B\text{Aut} \left(K \left(\pi_n \left(F \right), n \right) \right)$ to its first Postnikov truncation $B\text{Aut}_{\text{Grp}} \left(\pi_n \left(F \right) \right)$ is a section of the map g. The map s fits in yet another fiber sequence

$$K \left(\pi_n \left(F \right), n + 1 \right) \to B\text{Aut} \left(K \left(\pi_n \left(F \right), n \right) \right) \xrightarrow{s} B\text{Aut}_{\text{Grp}} \left(\pi_n \left(F \right) \right).$$

Thus, to give $\Psi_n \colon X \to B\text{Aut} \left(K \left(\pi_n \left(F \right), n \right) \right)$ up to homotopy is equivalent to:

1. An action of $\pi_1 \left(X \right)$ on $\pi_n \left(F \right)$ (by composition with s)
2. An element $o_n \in H^{n+1} \left(X, \pi_n \left(F \right) \right)$ with respect to this action.

The existence of a lift of Ψ_n along g, is equivalent to the vanishing of o_n. Moreover, the data that produces the lift is precisely a choice of a cochain representing o_n as a coboundary.

We can now put all of this together, denote $Y_n = \text{colim}_X P_n \circ \Phi$, by our assumption the fibres of the map $f \colon \colon Y \to X$ are connected and thus $Y_0 \cong X$. On the other hand we have $Y = \lim_n Y_n$. So, to give a section is the same as to give a compatible system of sections $s_n \colon X \to Y_n$. Now given a section $s_{n-1} \colon X \to Y_{n-1}$ the analysis

---

[1] For n=1 this statement requires that $\pi_1(F)$ is abelian, as we assumed as a simplifying assumption.

above gives an obstruction class $o_n \in H^{n+1}(X, \pi_n(F))$ that vanishes if and only it s_{n-1} can be lifted to a section $s_n : X \to Y_n$.

4.4.2 For ∞-Topoi and Linear(ized) Versions

The above discussion has a direct generalization to ∞-topoi and pro-objects in them. To give a general idea of how this works, let us consider the case where $X = \mathrm{Sh}(X)$ is the ∞-topos of sheaves of homotopy types on an actual topological space X. In this case, we can define Postnikov truncation of sheaves *level-wise*. The corresponding Eilenberg-MacLane objects are again just sheaves of Eilenberg-MacLane spaces. The ∞-category of pointed sheaves of Eilenberg-MacLane spaces of level n is equivalent to the ordinary category of sheaves of abelian groups on X (at least when $n \geq 2$). Applying π_n to a sheaf of homotopy types produces such sheaves of homotopy groups, and the resulting obstructions live in the *sheaf cohomology groups* $H^{n+1}(X, \pi_n F)$.

For an arbitrary map of ∞-topoi $f : X \to Y$, we can run the same obstruction theory and obtain a sequence of obstructions

$$o_n \in H^{n+1}(Y, \pi_n |X/Y|).$$

As with many situations in mathematics, it is sometimes easier to study a linearized version of the problem. In our case we can get linearized versions of the obstructions by considering sheafs of complexes instead of sheafs of spaces. In Sect. 4.2.7 we've discussed for every ring R, the functor

$$R \otimes - : S \to D_\infty(R).$$

Given a topological space X we can get a sheafified version of $R \otimes -$

$$R \otimes - : \mathrm{Sh}(X, S) \to \mathrm{Sh}(X, D_\infty(R))$$

This functor from ∞-sheaves of homotopy types on X to ∞-sheaves of complexes over R on X is obtained by applying $R \otimes -$ on each open set and then sheafifying. The ∞-category $\mathrm{Sh}(X, D_\infty(R))$ can also be direct ly constructed out of $\mathrm{Sh}(X, S)$ as we have

$$\mathrm{Sh}(X, D_\infty(R)) = \mathrm{Fun}^{\lim}(\mathrm{Sh}(X, S)^{op}, D_\infty(R)).$$

That is the ∞-category of contra-variant limit preserving functors from $\mathrm{Sh}(X, S)$ to $D_\infty(R)$.

For a general ∞-topos X we thus define:

$$X_R := \text{Fun}^{\lim}(X^{op}, \mathcal{D}_\infty(R)).$$

The ∞-cateogry X_R behaves now as the ∞-cateogry of homotopy sheaves of R-complexes over a "space". We also get a colimit preserving functor

$$R \otimes -: X \to X_R$$

we can now use these functors in order to linearize relative homotopy types. For a map of ∞-topoi $f: X \to Y$, We define the *relative R-homological shape of X over Y* to be

$$|X/Y|_R := R \otimes |X/Y| \in \text{Pro}(Y_R).$$

By truncating the homology above a certain degree, complexes admit an analogous (and simpler) obstruction theory to spaces. Applying this "homological" obstruction theory in our situation produces a sequence of obstructions of the form:

$$o_n^R \in H^{n+1}(Y, H_n(|X/Y|, R)).$$

While these obstruction are weaker from a theoretical point of view (we discarded some of the information), in practice, they tend to be much more computable as homology groups in general are much more computable than homotopy groups.

We can also linearise our objects using the universal stable ∞-cateogry Sp. By replacing $\mathcal{D}_\infty(R)$ with Sp everywhere in the discussion above we define:

$$X_{\mathbb{S}} := \text{Fun}^{\lim}(X^{op}, \text{Sp}).$$

and the *relative stable shape of X over Y* as

$$|X/Y|_{\mathbb{S}} := \Sigma_+^\infty |X/Y| \in \text{Pro}(Y_{\mathbb{S}}).$$

Furthermore we obtain a sequence of obstructions

$$o_n^{\mathbb{S}} \in H^{n+1}(Y, \pi_n^s |X/Y|),$$

where $\pi_n^s |X/Y|$ are the so-called *stable homotopy groups* of $|X/Y|$.

4.5 Étale Homotopy and Rational Points

4.5.1 The étale ∞-Topos

In the previous section we described an obstruction theory for the existence of sections of maps of (pro-)homotopy types constructed from ∞-topoi. We now want to apply this machinery to algebraic geometry. Let Sch be the category of schemes. The first step is to construct a functor

$$(-)_{\acute{e}t} : \mathrm{Sch} \to \mathrm{Topoi}_\infty.$$

We would like to mimic the construction that associates with a topological space X, the ∞-topoi of sheaves of homotopy types on X. The problem is that this uses the category of open subsets of X, and for schemes we do not have so many open subsets. The idea of Grothendieck was to use *local homeomorphisms* instead of actual open subsets. In algebraic geometry, local homeomorphisms correspond to *étale maps*.

Definition 16 For each scheme X, consider the full ∞-subcategory $\acute{e}t_{/X} \subseteq \mathrm{Sch}_{/X}$ spanned by étale maps $\varphi : Y \to X$. We define

$$X_{\acute{e}t} := \mathrm{Fun}\left(\acute{e}t_{/X}^{op}, \mathcal{S}\right)[\mathcal{W}^{-1}]$$

where $\mathcal{W}$ is the collection of morphisms $F \to G$ such that for every geometric point $x \in X$ the map

$$x^* F \to x^* G \in \mathcal{S}$$

is an equivalence. The ∞-category $X_{\acute{e}t}$ is an ∞-topos called the (small) *étale ∞-topos of* X. □

A map of schemes $f : X \to Y$, induces a map of ∞-topoi $f_{\acute{e}t} : X_{\acute{e}t} \to Y_{\acute{e}t}$. Thus, we can define the *relative étale shape* $|X_{\acute{e}t}/Y_{\acute{e}t}|$, which is an object of $\mathrm{Pro}\,(Y_{\acute{e}t})$. A section of f induces a global section of $|X_{\acute{e}t}/Y_{\acute{e}t}|$. Hence, our obstruction theory produces a sequence of elements

$$o_n \in H_{\acute{e}t}^{n+1}\left(Y, \pi_n |X_{\acute{e}t}/Y_{\acute{e}t}|\right)$$

whose vanishing is equivalent to the existence of a global section of $|X_{\acute{e}t}/Y_{\acute{e}t}|$, and the non-vanishing implies non-existence of sections of f.

4.5.2   Rational Points

An important special case is that of *rational points*. Let X be an algebraic variety over a field $\mathbb{F}$ and $X(\mathbb{F})$ the set of $\mathbb{F}$-points of X. A fundamental question is whether $X(\mathbb{F}) = \varnothing$. An $\mathbb{F}$-point of X is the same as a section of the structure morphism $t\colon X \to \mathrm{Spec}(\mathbb{F})$, hence we can apply the above theory for constructing obstructions to the existence of $\mathbb{F}$-points. First, let us spell out more explicitly where these obstructions live.

Write $\Gamma_{\mathbb{F}} = \mathrm{Gal}(\overline{\mathbb{F}}, \mathbb{F})$ for the absolute Galois (profinite) group of $\mathbb{F}$. For any profinite group Γ we can functorially associate an ∞-topoi $\mathcal{S}_{\Gamma}^{\mathrm{cts}}$. One way to define $\mathcal{S}_{\Gamma}^{\mathrm{cts}}$ is as follows: Let $\mathcal{S}et_{\Gamma}$ be the category of sets with continuous Γ-action. Then $\mathcal{S}_{\Gamma}^{\mathrm{cts}}$ is the ∞-category obtained by localizing the category $\mathcal{S}et_{\Gamma}^{\Delta^{op}}$ by Γ-equivariant maps $X \to Y$ such that the corresponding underlying map of simplicial sets is a weak-homotopy equivalence. We thus get from the Galois theory of fields

$$\mathrm{Spec}(\mathbb{F})_{\acute{e}t} \simeq \mathcal{S}_{\Gamma_{\mathbb{F}}}^{\mathrm{cts}}.$$

That is the ∞-category of homotopy types with a continuous action of the profinite group $\Gamma_{\mathbb{F}}$. The relative shape of X is an object of the ∞-category

$$\mathrm{Pro}\left(\mathrm{Spec}(\mathbb{F})_{\acute{e}t}\right) = \mathrm{Pro}\left(\mathcal{S}_{\Gamma_{\mathbb{F}}}^{\mathrm{cts}}\right).$$

Roughly, we have a pro-space $|X/\mathbb{F}|$ with a continuous action of the profinite Galois group $\Gamma_{\mathbb{F}}$. If we forget the action of $\Gamma_{\mathbb{F}}$, the mere pro-homotopy type of $|X/\mathbb{F}|$, can be identified with just the absolute étale pro-homotopy type of the base change of X to the algebraic closure $\overline{\mathbb{F}}$, denoted by $X_{\overline{\mathbb{F}}}$. In other words, we think of $X_{\overline{\mathbb{F}}}$ as the fiber of the map "$X \to B\Gamma_{\mathbb{F}}$". The group $\Gamma_{\mathbb{F}}$ acts on $X_{\overline{\mathbb{F}}}(\overline{\mathbb{F}})$ and the set of fixed points is exactly $X(\mathbb{F})$. On the other hand, the global sections of $|X/\mathbb{F}|$, are the *homotopy fixed points* of the action of $\Gamma_{\mathbb{F}}$ on $|X_{\overline{\mathbb{F}}}|$.

Definition 17 We denote by $X(\mathbb{F}^h)$ the π_0 of the space of homotopy fixed points (i.e. global sections) of $\Gamma_{\mathbb{F}}$ on $|X_{\mathbb{F}}| = |X/\mathbb{F}|$ and call it the *homotopy obstruction set*.   □

The name comes from the fact that we have a canonical map of sets

$$X(\mathbb{F}) \to X(\mathbb{F}^h)$$

and therefore if $X(\mathbb{F}^h) = \varnothing$, then also $X(\mathbb{F}) = \varnothing$. There are special cases in which this map recovers some classical notions, that can be phrased without the language of homotopy theory.

Example 4 Let E be an elliptic curve over $\mathbb{Q}$. The relative étale homotopy type of E over $\mathbb{Q}$ is equivalent to a profinite torus $\hat{\mathbb{T}}$, which is the classifying space of the *Tate module* of E. This can be written as a formal inverse limit as follows:

$$|E/\mathbb{Q}| \simeq \text{``}\lim_{n \in \mathbb{N}^\times} B\left(E[n]\right)\text{''}.$$

In this case, the map $E(\mathbb{Q}) \to E(\mathbb{Q}^h)$ can be identified with the classical *Kummer map*.

For the question of whether $X(\mathbb{F}^h) = \varnothing$, the obstruction theory outlined above, specializes in this case to give a sequence of obstructions in (continuous) Galois cohomology:

$$o_n \in H^{n+1}\left(\Gamma_{\mathbb{F}}, \pi_n |X/\mathbb{F}|\right) \simeq H_{\acute{e}t}^{n+1}\left(\Gamma_{\mathbb{F}}, \pi_n^{\acute{e}t}(X_{\bar{\mathbb{F}}})\right).$$

The case $n = 1$ has a classical interpretation. The vanishing of

$$o_1 \in H^2\left(\Gamma_{\mathbb{F}}, \pi_1^{\acute{e}t}(X_{\bar{\mathbb{F}}})\right)$$

is equivalent to the existence of a section to the map of groups

$$\pi_1^{\acute{e}t}(X) \to \pi_1^{\acute{e}t}(\text{Spec}(\mathbb{F})) \simeq \Gamma_{\mathbb{F}},$$

which is the classical *Grothendieck section obstruction* . Furthermore, if by chance $\pi_n^{\acute{e}t}(X) = 0$ for all $n \geq 2$ (e.g. a curve of positive genus) we can identify the set $X(\mathbb{F}^h)$ with the set of sections up to conjugation of the short exact sequence

$$1 \to \pi_1^{\acute{e}t}(X_{\bar{\mathbb{F}}}) \to \pi_1^{\acute{e}t}(X) \to \Gamma_{\mathbb{F}} \to 1.$$

It is a famous conjecture by Grothendieck that if $\mathbb{F}$ is a number field and X is smooth and proper curve of genus bigger than one, then the map

$$X(\mathbb{F}) \to X(\mathbb{F}^h)$$

is a bijection.

Recall that we also defined linearized versions of the obstruction theory using the *homological shape* or *stable shape*. We use the notation $X(\mathbb{F}^{h\mathbb{Z}})$ and $X(\mathbb{F}^{h\mathbb{S}})$ for the π_0 of the space of sections of the maps

$$|X/\mathbb{F}|_{\mathbb{Z}} \to |\mathbb{F}/\mathbb{F}|_{\mathbb{Z}} \in \text{Pro}((\mathbb{F}_{\acute{e}t})_{\mathbb{Z}})$$

and

$$|X/\mathbb{F}|_{\mathbb{S}} \to |\mathbb{F}/\mathbb{F}|_{\mathbb{S}} \in \text{Pro}((\mathbb{F}_{\acute{e}t})_{\mathbb{S}})$$

respectively. Observe that we have a hierarchy of *strength* for the obstruction sets given by the factorizations of maps:

$$X(\mathbb{F}) \to X(\mathbb{F}^h) \to X(\mathbb{F}^{h\mathbb{S}}) \to X(\mathbb{F}^{h\mathbb{Z}}).$$

It turns out that a homotopy section for these linear versions can be produced not only by a fixed point, but also by a certain *linear version* of it.

Definition 18 The free abelian group $\mathbb{Z}X(\bar{\mathbb{F}})$ on the set $X(\bar{\mathbb{F}})$ carries a $\Gamma_{\mathbb{F}}$-action given by linear extension of the action of $\Gamma_{\mathbb{F}}$ on $X(\bar{\mathbb{F}})$. A *zero-cycle* on X is defined to be an element of the fixed points $\left(\mathbb{Z}X(\bar{\mathbb{F}})\right)^{\Gamma_{\mathbb{F}}}$. The abelian group $\mathbb{Z}X(\bar{\mathbb{F}})$ also admits a natural degree map

$$\deg: \mathbb{Z}X(\bar{\mathbb{F}}) \to \mathbb{Z}$$

sending a formal sum to the sum of its coefficients. Zero-cycles of degree one are in particular a linear analogue, and a strict generalization, of a point in $X(\mathbb{F}) = X(\bar{\mathbb{F}})^{h\Gamma_{\mathbb{F}}}$.

It turns out that every zero-cycle of degree one induces an element of $X(\mathbb{F}^{h\mathbb{S}})$ and hence also of $X(\mathbb{F}^{h\mathbb{Z}})$. Thus, the associated obstructions, obstruct the existence of zero-cycles of degree one and not only actual rational points.   □

Of course, we can also use homology with coefficient in other rings than $\mathbb{Z}$ for computing obstructions. The following is an explicit example with coefficients in $\mathbb{F}_2$:

Example 5 (Arad–Carmeli–S, see [1]) Let $\mathbb{F}$ be a field of characteristic different than 2 and let $a_i \in \mathbb{F}^{\times}$ for $0 \leq i \leq n$. Let

$$X = V\left(\sum_{i=0}^n a_i x_i^2 = 1\right) \subseteq \mathbb{A}^{n+1}$$

be a quadric. We have

$$\tilde{H}_k^{\acute{e}t}(X_{\bar{\mathbb{F}}}; \mathbb{F}_2) \cong \begin{cases} \mathbb{F}_2 & k = n \\ 0 & else \end{cases}.$$

The $\mathbb{F}_2$-homology obstruction is therefore a class $o_n^{\mathbb{F}_2} \in H_{Gal}^{n+1}(\mathbb{F}, \mathbb{F}_2)$. This class can be computed explicitly as the cup product

$$o_n^{\mathbb{F}_2} = \cup_{i=0}^n (a_i),$$

where

$$(a_i) \in H^1(\mathrm{Spec}(\mathbb{F}), \mathbb{F}_2) \cong \mathbb{F}^{\times}/(\mathbb{F}^{\times})^2.$$

4.5.3 The Local-to-Global Principle

When working over a *number field* one can obtain another type of obstructions for rational points using the *local-to-global principle*. For simplicity, let us assume that $\mathbb{F} = \mathbb{Q}$ and that X is smooth and proper over $\mathbb{Q}$. For each prime number p, we have the field $\mathbb{Q}_p$ of p-adic numbers and by functoriality we obtain a map of sets

$$X(\mathbb{Q}) \to \prod_{p \leq \infty} X(\mathbb{Q}_p) = X(\mathbb{A})$$

where $\mathbb{Q}_\infty = \mathbb{R}$. Thus, the set $X(\mathbb{A})$ of adelic points (which is usually much easier to compute) is an obstruction set for rational points in the sense that if $X(\mathbb{A}) = \emptyset$ then $X(\mathbb{Q}) = \emptyset$. It is a classical theorem of Minkowski if X is a quadric hypersurface we have

$$X(\mathbb{A}) \neq \emptyset \Rightarrow X(\mathbb{Q}) \neq \emptyset.$$

However for general varieties, by itself, this is not a complete obstruction. There are many example for X such that $X(\mathbb{A}) \neq 0$ and still $X(\mathbb{Q}) = \emptyset$. Thus, one would like to further restrict $X(\mathbb{A})$ in a certain way. Manin [13] originated the classical method called the *Brauer-Manin obstruction* that goes as follows. Let $\mathrm{Br}(X) = H^2_{\acute{e}t}(X, \mathbb{G}_m)$ be the Brauer group of X. There is a natural map

$$\omega \colon X(\mathbb{A}) \times \mathrm{Br}(X) \to \bigoplus_p H^2(\mathbb{Q}_p, \mathbb{G}_m) \xrightarrow{\Sigma(\mathrm{inv}_p)} \mathbb{Q}/\mathbb{Z}$$

such that a point $x \in X(\mathbb{A})$, that comes from a rational point, must satisfy $\omega(x, g) = 0$ for all $g \in \mathrm{Br}(X)$. Let us denote by $X(\mathbb{A})^{\mathrm{Br}} \subseteq X(\mathbb{A})$ the set of all such points. It follows that if $X(\mathbb{A})^{\mathrm{Br}} = \emptyset$, then $X(\mathbb{Q}) = \emptyset$.

This algebraically defined obstruction set can in fact be also constructed using the *homotopical methods* discussed above, particularly using the homological obstruction sets. Consider the commutative diagram

$$
\begin{array}{ccc}
X(\mathbb{Q}) & \longrightarrow & X(\mathbb{Q}^{h\mathbb{Z}}) \\
\downarrow & & \downarrow \\
\prod_{p \leq \infty} X(\mathbb{Q}_p) & \longrightarrow & \prod_{p \leq \infty} X(\mathbb{Q}_p^{h\mathbb{Z}})
\end{array}
$$

We have $X(\mathbb{A}) = \prod_{p \leq \infty} X(\mathbb{Q}_p)$ and we denote $X(\mathbb{A}^{h\mathbb{Z}}) := \prod_{p \leq \infty} X(\mathbb{Q}_p^{h\mathbb{Z}})$. This induces a factorization

$$X(\mathbb{Q}) \to X(\mathbb{A}) \times_{X(\mathbb{A}^{h\mathbb{Z}})} X(\mathbb{Q}^{h\mathbb{Z}}) \to X(\mathbb{A}).$$

Letting $X(\mathbb{A})^{h\mathbb{Z}} \subseteq X(\mathbb{A})$ be the image of the second map, we obtain again a (possibly) smaller obstruction set then $X(\mathbb{A})$.

Theorem 19 (Harpaz–S, see [8]) $X(\mathbb{A})^{h\mathbb{Z}} = X(\mathbb{A})^{\mathrm{Br}}$. $\qquad\qquad\qquad\qquad\qquad\square$

Proof (Proof (sketch)) First we give a different description of $X(\mathbb{A})^{\mathrm{Br}}$. Let $\mu_\infty \subset \mathbb{G}_m$ be the étale sheaf of roots of unity. It can be shown that the map

$$H_{\acute{e}t}^2(X, \mu_\infty) \to H_{\acute{e}t}^2(X, \mathbb{G}_m) = \mathrm{Br}(X)$$

is surjective for X smooth and an isomorphism when X is a spectrum of a field. Thus, we can define the set $X(\mathbb{A})^{\mathrm{Br}}$ using $H_{\acute{e}t}^2(X, \mu_\infty)$ instead of $\mathrm{Br}(X)$ and get the same set.

Now, recall that $X(\mathbb{Q}^{h\mathbb{Z}})$ is defined as the π_0 of the space of sections of the map

$$|X/\mathbb{Q}|_{\mathbb{Z}} \to |\mathbb{Q}/\mathbb{Q}|_{\mathbb{Z}}$$

in the category $\mathrm{Pro}((\mathbb{Q}_{\acute{e}t})_{\mathbb{Z}})$. More explicitly, $|\mathbb{Q}/\mathbb{Q}|_{\mathbb{Z}}$ is just the constant sheaf $\mathbb{Z}$. So $X(\mathbb{Q}^{h\mathbb{Z}}) = \pi_0(F)$ where F is the fiber of the map of spaces

$$\mathrm{Map}_{\mathrm{Pro}((\mathbb{Q}_{\acute{e}t})_{\mathbb{Z}})}(\mathbb{Z}, |X/\mathbb{Q}|_{\mathbb{Z}}) \to \mathrm{Map}_{\mathrm{Pro}((\mathbb{Q}_{\acute{e}t})_{\mathbb{Z}})}(\mathbb{Z}, \mathbb{Z}),$$

over the identity map $\mathrm{Id}_{\mathbb{Z}}$. Comparing with the local version we get the following interpretation of $X(\mathbb{A})^{h\mathbb{Z}}$: A point $x_0 \in X(\mathbb{A})$ belongs to $X(\mathbb{A})^{h\mathbb{Z}}$ if and only if its image in[2]

$$\pi_0\left(\mathrm{Map}_{\mathrm{Pro}((\mathbb{A}_{\acute{e}t})_{\mathbb{Z}})}(\mathbb{Z}, |X/\mathbb{Q}|_{\mathbb{Z}})\right)$$

is in the image of the localisation map

$$\pi_0\left(\mathrm{Map}_{\mathrm{Pro}((\mathbb{Q}_{\acute{e}t})_{\mathbb{Z}})}(\mathbb{Z}, |X/\mathbb{Q}|_{\mathbb{Z}})\right) \to \pi_0\left(\mathrm{Map}_{\mathrm{Pro}((\mathbb{A}_{\acute{e}t})_{\mathbb{Z}})}(\mathbb{Z}, |X/\mathbb{Q}|_{\mathbb{Z}})\right).$$

In general an object $A \in \mathrm{Pro}((\mathbb{Q}_{\acute{e}t})_{\mathbb{Z}})$ is a pro-system of complexes of Galois $\Gamma_{\mathbb{Q}}$-modules and

$$\pi_i(\mathrm{Map}_{\mathrm{Pro}((\mathbb{Q}_{\acute{e}t})_{\mathbb{Z}})}(\mathbb{Z}, A)) = \mathbb{H}^{-i}(\mathbb{Q}, A).$$

---

[2]The suitable definition of the ∞-topos $\mathbb{A}_{\acute{e}t}$ requires some care, we shall not elaborate on this issue.

Consider now the map

$$H^0_{\acute{e}t}(\mathbb{Q}; |X_{\acute{e}t}/\mathbb{Q}|_{\mathbb{Z}}) \to H^0_{\acute{e}t}(\mathbb{A}; |X_{\acute{e}t}/\mathbb{Q}|_{\mathbb{Z}})$$

We can try to understand when for a general finite module M of the Galois group of $\mathbb{Q}$ and an element of $H^i_{\acute{e}t}(\mathbb{A}; M)$, we have a lift to $H^i_{\acute{e}t}(\mathbb{Q}; M)$. Denote by A^* the Pontryagin dual of A and recall the Poitou-Tate exact sequence :

$$0 \to H^0_{\acute{e}t}(\mathbb{Q}; M) \to H^0_{\acute{e}t}(\mathbb{A}; M) \xrightarrow{\beta} H^2_{\acute{e}t}(\mathbb{Q}; \hom(M, \mu_\infty))^* \to$$

$$\to H^1_{\acute{e}t}(\mathbb{Q}; M) \to H^1_{\acute{e}t}(\mathbb{A}; M) \to H^1_{\acute{e}t}(\mathbb{Q}; \hom(M, \mu_\infty))^* \to$$

$$\to H^2_{\acute{e}t}(\mathbb{Q}; M) \to H^2_{\acute{e}t}(\mathbb{A}; M) \to H^0_{\acute{e}t}(\mathbb{Q}; \hom(M, \mu_\infty))^* \to 0.$$

the map β can be described by the pairing

$$\beta^*: H^i_{\acute{e}t}(\mathbb{A}; M) \otimes H^{2-i}_{\acute{e}t}(\mathbb{Q}; \hom(M, \mu_\infty)) \to \mathbb{Q}/\mathbb{Z}.$$

Thus the exactness of the Poitou-Tate sequence implies that the condition for a class in $H^i_{\acute{e}t}(\mathbb{A}; M)$ to come from a class in $H^i_{\acute{e}t}(\mathbb{Q}; M)$ is precisely to be in the kernel of β^*. The analogous statement remains true for a **pro-complex** of modules like $|X_{\acute{e}t}/\mathbb{Q}|_{\mathbb{Z}}$. It remains to identify β^* for the relevant i with the Brauer-Manin pairing. A simple diagram chase with adjoints gives

$$|X_{\acute{e}t}/\mathbb{Q}|_{\mathbb{Z}} = t^{\mathbb{Z}}_! \mathbb{Z}$$

where $t: X \to \mathrm{Spec}\,(\mathbb{Q})$ is the structure map and

$$t^{\mathbb{Z}}_!: \mathrm{Pro}((X_{\acute{e}t})_{\mathbb{Z}}) \to \mathrm{Pro}((\mathbb{Q}_{\acute{e}t})_{\mathbb{Z}})$$

is the pro-left adjoint to the pullback functor

$$t^*_{\mathbb{Z}}: (\mathbb{Q}_{\acute{e}t})_{\mathbb{Z}} \to (X_{\acute{e}t})_{\mathbb{Z}}.$$

We now observe that

$$H^2_{\acute{e}t}(\mathbb{Q}, \hom(|X_{\acute{e}t}/\mathbb{Q}|, \mu_\infty)) \cong [\Sigma^{-2}\mathbb{Z}, \hom(t_!\mathbb{Z}, \mu_\infty)] = [\Sigma^{-2}\mathbb{Z}, \hom(\mathbb{Z}, t^*\mu_\infty)] = H^2_{\acute{e}t}(X, \mu_\infty).$$

The proof is completed by identifying the two pairings by a simple diagram chase.

$\square$

Of course, one can obtain a stronger obstruction set if one uses the above technique with the p-adic *homotopy* obstruction sets instead of the *homology* ones. Namely, using the diagram

$$
\begin{array}{ccc}
X(\mathbb{Q}) & \longrightarrow & X(\mathbb{Q}^h) \\
\downarrow & & \downarrow \\
\prod_{p \leq \infty} X(\mathbb{Q}_p) & \longrightarrow & \prod_{p \leq \infty} X(\mathbb{Q}_p^h)
\end{array}
$$

Denote $X(\mathbb{A})^h \subseteq X(\mathbb{A})$ the corresponding image of the projection from the pullback (as above). This obstruction set has also a more classical description.

Theorem 20 (Harpaz–S, see [8]) $X(\mathbb{A})^h$ *is the étale-Brauer-Manin obstruction set defined by Skorobogatov in [15].*   □

Finally, one might also wonder what we get using the intermediate *stable* obstruction sets. It turns out that the resulting algebraic obstructions can be written down explicitly using a spectral version of Poitou-Tate theory in which, among other things, μ_∞ is replaced by a twisted version of the Brown-Comenetz spectrum. This is a work in progress with Vesna Stojanoska.

4.6   Galois Theory and Embedding Problems

Similar homotopical obstructions can also be used for problems in Galois theory. One kind of Galois theoretic problems amenable to homotopical methods are "embedding problems". Let Γ be a profinite group. A finite embedding problem **E** for Γ is a diagram

$$
\begin{array}{ccc}
 & & \Gamma \\
 & & \downarrow p \\
G & \xrightarrow{\;f\;} & H
\end{array}
$$

where G and H are finite groups, f is surjective and p is continuous for H with the discrete topology. We say that the embedding problem has a proper solution if there exists a continuous surjective homomorphism $q : \Gamma \to G$ such that $p = fq$. The embedding problem has a weak solution if there exists a continuous map $q : \Gamma \to G$ (not neccesarily surjective) such that $p = fq$. We will call $\mathrm{Ker}(\mathbf{E}) := \mathrm{Ker}(f)$ the kernel of the embedding problem **E**. The relevance of this definition to Galois theory becomes apparent when one takes Γ to be the Galois group of some field extension.

Indeed, let F be a field and $K/F \subset E/F$ be a pair of Galois extensions of F. Assume that we are given a surjective map

$$G \xrightarrow{\ f\ } \mathrm{Gal}\,(K/F).$$

By the Galois correspondence theorem, giving a Galois extension $K/F \subset L/F \subset E/F$ with an isomorphism $G \cong \mathrm{Gal}\,(L/F)$ and an identification of f with a restriction map $\mathrm{Gal}\,(L/F) \to \mathrm{Gal}\,(K/F)$ is the same as giving a proper solution to the embedding problem

$$\begin{array}{c} \mathrm{Gal}\,(E/F) \\ \Big\downarrow{\scriptstyle p} \\ G \xrightarrow{\ f\ } \mathrm{Gal}\,(K/F). \end{array}$$

Thus, embedding problems arise naturally when trying to construct Galois extensions inductively.

4.6.1 Topoi and Embedding Problems

Similar to the case of rational points, we introduce homotopical obstructions for embedding problems by first translating the problem of having a lift $q : \Gamma \to G$ to the existence of a section for some map between ∞-topoi. Consider the pullback profinite group

$$\begin{array}{ccc} \Gamma(\mathbf{E}) := \Gamma \times_H G & \xrightarrow{\ g\ } & \Gamma \\ \Big\downarrow & & \Big\downarrow{\scriptstyle p} \\ G & \xrightarrow[\ f\ ]{} & H. \end{array}$$

It is easy to see that having a weak solution to $\mathbf{E}$ is equivalent to having a section for the projection map $g : \Gamma(\mathbf{E}) \to \Gamma$. The map g gives a geometric morphism:

$$\mathcal{S}^{\mathrm{cts}}_{\Gamma(\mathbf{E})} \overset{g^*}{\underset{g_*}{\rightleftarrows}} \mathcal{S}^{\mathrm{cts}}_{\Gamma}.$$

Denote the relative shape

$$X_{\mathbf{E}} = \left| \mathcal{S}^{\mathrm{cts}}_{\Gamma(\mathbf{E})} / \mathcal{S}^{\mathrm{cts}}_{\Gamma} \right| \in \mathrm{Pro}(\mathcal{S}^{\mathrm{cts}}_{\Gamma}).$$

Direct computation shows that $X_{\mathbf{E}}$ lies in the essential image $\mathcal{S}^{cts} \subset \mathrm{Pro}(\mathcal{S}_\Gamma^{cts})$ and that it's underlying homotopy type is of the Eilenberg-MacLane space $K(\mathrm{Ker}(\mathbf{E}), 1)$.

Thus, the only obstruction we get this way is

$$o_1 \in H^2(\Gamma, \pi_1(X_{\mathbf{E}})) = H^2(\Gamma, \mathrm{Ker}(\mathbf{E})).$$

The obstruction o_1 is not helpful, as is turns out to be just a reformulation of the original problem. However, taking homological obstructions can give new results. Indeed, for any ring R we get a sequence of obstructions to the existence of a weak solution

$$o_n^R \in H^{n+1}(\Gamma, H_n(X_{\mathbf{E}}, R)) = H^{n+1}(\Gamma, H_n(\mathrm{Ker}(\mathbf{E}), R)).$$

The obstruction $o_1^{\mathbb{Z}} \in H^2(\Gamma, H_1(\mathrm{Ker}(\mathbf{E}), \mathbb{Z})) = H^2(\Gamma, \mathrm{Ker}(\mathbf{E})^{ab})$ is equivalent to the existence of a weak solution for the embedding problem:

$$
\begin{array}{ccc}
 & & \Gamma \\
 & & \downarrow{\scriptstyle p} \\
G/[\mathrm{Ker}(\mathbf{E}), \mathrm{Ker}(\mathbf{E})] & \xrightarrow{\ f\ } & H,
\end{array}
$$

and was considered classically. However, the obstructions o_n^R for $n > 1$ can be used even for embedding problems with perfect kernel. For example, if q is an odd prime power, the group $K = \mathrm{PSL}(2, q^2)$ has $H_1(K, \mathbb{Z}) = 0$ and $H_2(K, \mathbb{Z}) \neq 0$. Analysing embedding problems with $G \to H$ being $\mathrm{Aut}(\mathrm{PSL}(2, q^2)) \to \mathrm{Out}(\mathrm{PSL}(2, q^2))$ and using non-trivial $o_2^{\mathbb{Z}}$ obstructions we can show for example:

Theorem 21 (Carlson-S [6]) *Let p_1, p_2, p_3 be three primes such that*

$$p_1 p_2 p_3 \equiv 3 \bmod 4$$

and

$$\left(\frac{p_i}{p_j}\right) = -1$$

for all $i \neq j$. , and let $K = \mathbb{Q}(\sqrt{-p_1 p_2 p_3})$. Then $\mathrm{Aut}(\mathrm{PSL}(2, q^2))$ cannot be realized as the Galois group of an unramified extension of K while the maximal solvable quotient $\mathrm{Aut}(\mathrm{PSL}(2, q^2))^{sol} \cong \mathrm{Out}(\mathrm{PSL}(2, q^2)) \cong \mathbb{Z}/2 \times \mathbb{Z}/2$ can be realized.   □

Dirichlet's Theorem implies that there are infinitely many such triples of primes (for example $3, 5, 17$).

Acknowledgments The text is based on the talks given by the author in the LMS Research School on "Homotopy Theory and Arithmetic Geometry: Motivic and Diophantine Aspects" held in imperial college in July 2018. I wish to thank Frank Neumann and Ambrus Pál for organising this outstanding event and inviting me to speak in it. I would like to thank Shay Ben Moshe for his remarks on the first draft. Finally, I would also like to thank Shachar Carmeli and Lior Yanovski for their indispensable help with the preparation of this notes.

References

1. E. Arad, S. Carmeli, T.M. Schlank, étale homotopy obstructions of arithmetic spheres (2019). arXiv preprint
2. M. Artin, B. Mazur, Etale homotopy, in *Lecture Notes in Mathematics*, vol. 100 (Springer, Berlin, 2006)
3. C. Barwick, D.M. Kan, Relative categories: another model for the homotopy theory of homotopy theories. Indag. Math. **23**(1–2), 42–68 (2012)
4. J.E. Bergner, Three models for the homotopy theory of homotopy theories. Topology **46**(4), 397–436 (2007)
5. J.E. Bergner, A survey of $(\infty, 1)$-categories, in *Towards Higher Categories* (2010), pp. 69–83
6. M. Carlson, T.M. Schlank, The unramified inverse galois problem and cohomology rings of totally imaginary number fields (2016). arXiv preprint
7. E.M. Friedlander, Etale homotopy of simplicial schemes (AM-104), in *Annals of Mathematical Studies*, vol. 104 (Princeton University, Princeton, 1982)
8. Y. Harpaz, T.M. Schlank, Homotopy obstructions to rational points (2011). arXiv preprint
9. A. Joyal, Quasi-categories and kan complexes. J. Pure Appl. Algebra **175**(1–3), 207–222 (2002)
10. A. Joyal, *Notes on Quasi-categories* (2008). preprint
11. J. Lurie, Higher topos theory, in *Annals of Mathematics Studies*, vol. 170 (Princeton University, Princeton, 2009)
12. J. Lurie, *Higher Algebra* (2012)
13. Y.I. Manin, *Le groupe de Brauer–Grothendieck en géométrie diophantienne* (1971)
14. C. Rezk, A model for the homotopy theory of homotopy theory. Trans. Am. Math. Soc. **353**(3), 973–1007 (2001)
15. A.N. Skorobogatov, Beyond the Manin obstruction. Invent. Math. **135**, 399–424 (1999)

Chapter 5
$\mathbb{A}^1$-homotopy Theory and Contractible Varieties: A Survey

Aravind Asok and Paul Arne Østvær

Abstract We survey some topics in $\mathbb{A}^1$-homotopy theory. Our main goal is to highlight the interplay between $\mathbb{A}^1$-homotopy theory and affine algebraic geometry, focusing on the varieties that are "contractible" from various standpoints.

5.1 Introduction: Topological and Algebro-Geometric Motivations

In this note, we want to survey the theory of varieties that are "weakly contractible" from the standpoint of the Morel–Voevodsky $\mathbb{A}^1$-homotopy theory of algebraic varieties. The current version of this document is based on lectures given by the first author at the Fields Institute workshop entitled "Group actions, generalized cohomology theories and affine algebraic geometry" at the University of Ottawa in 2011, and the Nelder Fellow Lecture Series by the second author in connection with the research school "Homotopy Theory and Arithmetic Geometry: Motivic and Diophantine Aspects" at Imperial College London in 2018.

As a source of inspiration, we review some aspects of the theory of open contractible manifolds, which has source in one of the first false proofs of the Poincaré conjecture, and served as a testing ground for many ideas of classical geometric topology in the 1950s and 1960s. We also review some aspects of the theory of complex algebraic varieties whose associated complex manifolds are contractible. Along the way we formulate some problems that we feel are interesting.

A. Asok
Department of Mathematics, University of Southern California, Los Angeles, CA, USA
e-mail: asok@usc.edu

P. A. Østvær (✉)
Department of Mathematics "F. Enriques", University of Milan, Milan, Italy

Department of Mathematics, University of Oslo, Oslo, Norway
e-mail: paularne@math.uio.no

F. Neumann, A. Pál (eds.), *Homotopy Theory and Arithmetic Geometry – Motivic and Diophantine Aspects*, Lecture Notes in Mathematics 2292,
https://doi.org/10.1007/978-3-030-78977-0_5

145

5.1.1 Open Contractible Manifolds

Suppose M^n is an n-dimensional manifold, which we take to mean either *topological, piecewise-linear (PL)* or *smooth* (we will specify). Say an n-manifold M^n is open if it is non-compact and without boundary. As usual, let I be the unit interval $[0, 1]$. Recall that M^n is called *contractible* if it is homotopy equivalent to a point. Euclidean space $\mathbb{R}^n$ is the primordial example of an open contractible manifold: radially contract every point to the origin.

The history of open contractible manifolds is closely intertwined with the history of geometric topology. To explain this connection, recall that the classical *Poincaré conjecture* asks whether every closed manifold M^n homotopy equivalent to S^n is actually homeomorphic to S^n. The history of the Poincaré conjecture is closely connected with the history of open contractible manifolds.

In 1934, J.H.C. Whitehead gave what he thought was a proof of the Poincaré conjecture in dimension 3 [149]. In brief, his argument went as follows: start with a homotopy equivalence $f : M^3 \longrightarrow S^3$. Removing a point produces an open contractible manifold $M^3 \setminus pt$ and continuous map $f : M^3 \setminus pt \to \mathbb{R}^3$. In essence, Whitehead argued that all open contractible 3-manifolds are homeomorphic to $\mathbb{R}^3$. If this was true, then such a homeomorphism would extend by continuity across infinity inducing a homemorphism $M^3 \to S^3$. Unfortunately, this proof collapsed soon thereafter as Whitehead constructed an open contractible manifold $\mathcal{W}$ (the so-called Whitehead manifold) that is not homeomorphic to the Euclidean space $\mathbb{R}^3$ [150]. Whitehead's example raised the question of characterizing Euclidean space among all open contractible manifolds, perhaps in some homotopic way.

Whitehead's construction is delightfully geometric and is given as an open subset of a Euclidean space with closed complement the "Whitehead continuuum", itself built out of intricately linked tori. Since the construction of the Whitehead manifold is relatively simple, we give it here. Consider the Whitehead link in S^3; thicken it to obtain two linked solid tori; label the interior solid torus $\hat{T}_1$ and the exterior solid torus T_0. The complement of each T_i in S^3 is unknotted. Since $\hat{T}_1$ is unknotted, the complement T_1 of the interior of $\hat{T}_1$ in S^3 is another unknotted solid torus that contains T_0. Choose a homeomorphism h of S^3 that maps T_0 onto T_1. Proceeding inductively, may therefore construct solid tori $T_0 \subset T_1 \subset \cdots$ in S^3 by setting $T_{j+1} = h(T_j)$. The union $\mathcal{W} := \cup_i T_i$ is the required open contractible manifold. To see that $\mathcal{W}$ is simply connected, one proceeds as follows. Observe that every closed loop in M^3 is contained in some T^j by construction. Now, observe that every closed loop in T_0 may be shrunk to a point (possibly crossing through itself) in T^1, and thus every closed loop in T_j may be shrunk to a point in T_{j+1}.

Whitehead's original proof that $\mathcal{W}$ is not homeomorphic to $\mathbb{R}^3$ is essentially geometric. However, it was later observed that for any sufficiently large compact subset $K \subset \mathcal{W}$ (K contains T_0 suffices), the complement $\mathcal{W} \setminus K$ is not simply connected. More generally, suppose M is an open contractible manifold. We may consider the collection of compact sets $K \subset M$ ordered with respect to inclusion. In good situations, the inverse system $\pi_1(M \setminus K)$ stabilizes and defines an invariant of the homeomorphism type of M called the fundamental group at infinity and denoted $\pi_1^\infty(M)$. Observe that $\mathbb{R}^3$ is simply-connected at infinity, since any compact subset is contained in a sufficiently large ball, whose complement in $\mathbb{R}^3$ is simply connected.

While Whitehead's construction lay dormant for some time, open contractible manifolds were studied with great intensity in the late 1950s and 1960s. Although $\mathcal{W}$ is not homeomorphic to $\mathbb{R}^3$ Glimm and Shapiro pointed out that the cartesian product $\mathcal{W} \times \mathbb{R}^1$ is homeomorphic to $\mathbb{R}^4$ [58], i.e., $\mathcal{W}$ does not have the *cancellation property* with respect to Cartesian products with Euclidean spaces. Moreover, Glimm showed that the self-product $\mathcal{W} \times \mathcal{W}$ is homeomorphic to $\mathbb{R}^6$ [58]. These kinds of "non-cancellation" phenomenon became another theme explored in the study of open contractible manifolds.

By modifying the construction of the Whitehead continuum, D.R. McMillan constructed infinitely many pairwise non-homeomorphic open contractible 3-manifolds [100]; each of these 3-manifolds could be embedded in $\mathbb{R}^3$. While McMillan's original proof that his manifolds were non-homeomorphic was essentially geometric, it was observed that his examples actually have distinct fundamental groups at infinity. This naturally raises the question of which groups may appear as a fundamental groups at infinity of open contractible 3-manifolds.

Simultaneously, McMillan showed his examples also had the property that taking a Cartesian product with the real line yielded a manifold homeomorphic to $\mathbb{R}^4$. In fact, McMillan showed that the product of *any* open contractible 3-manifold M and the real line was homeomorphic to $\mathbb{R}^4$ assuming the 3-dimensional Poincaré conjecture. Later joint work of J. Kister and McMillan [87] established existence of open contractible 3-manifolds that could not be embedded in $\mathbb{R}^3$. Work of Zeeman and McMillan then showed that the product of any open contractible 3-manifold and $\mathbb{R}^2$ was homeomorphic to $\mathbb{R}^5$. From one point of view, the zoo of open contractible manifolds presented some kind of "lower bound" on the complexity of the Poincaré conjecture.

The existence of open contractible manifolds of higher dimensions was also studied in conjunction with the higher dimensional Poincaré conjecture. Constructions of open contractible 4-manifolds were initially given by Mazur and Poenaru [99, 115], but as with many other constructions in geometric topology, these isolated examples in dimension 4 did not at first fit into a general picture. Dimension ≥ 5 proved itself to be more tractable. Extending the results of McMillan, for any integer $n \geq 5$, Curtis and Kwun [35] constructed uncountably many pairwise non-homeomorphic open contractible n-manifolds.

The method of Curtis and Kwun in some sense mirrors McMillan's construction, but requires significantly more group theoretic input. The basic idea is to construct open contractible manifolds with a prescribed (finitely presented) fundamental group at infinity. If P is a finitely presented group, then one may build many cell complexes $K(P)$ with fundamental group P. If one embeds $K(P)$ in S^n for $n \geq 5$, then the boundary of a regular neighborhood N of $K(P)$ has fundamental group P (this fails for $n = 4$, since the fundamental group of the boundary depends on the embedding). One says that $K(P)$ is a homologically trivial presentation if $H_i(K(P), \mathbb{Z})$ is trivial for $i = 1, 2$. In particular, if P admits a homologically trivial presentation then the Hurewicz theorem implies that $K(P)$ is a perfect group. If N is a regular neighborhood of an embedding $K(P) \hookrightarrow S^n$, $n \geq 5$, then one sees that $S^n \setminus \partial N$ is contractible if $K(P)$ is a homologically trivial presentation, but has fundamental group at infinity P. The output is that one has countably many compact contractible manifolds that act as "building blocks" for an uncountable collection by taking suitable connected sums. The method of Curtis and Kwun for producing "building blocks" evidently does not extend to $n = 4$. Later, Glaser extended Mazur's construction to produce countably many "building blocks" from which uncountably many pairwise non-homeomorphic open contractible 4-manifolds could be built by taking suitable connected sums as above [57]. We summarize these results in the following statement.

Theorem 1 *In every dimension ≥ 3, there exists uncountably many pairwise non homeomorphic open contractible n-manifolds.* $\qquad\qquad\square$

Returning to non-cancellation phenomena, Stallings [132] generalized the results of Zeeman and McMillan by observing:

1. for any integer $n \geq 5$, $\mathbb{R}^n$ is the unique open contractible PL-manifold that is simply-connected at infinity; and
2. if M^n is an open contractible m-manifold, then $M^n \times \mathbb{R}$ is always simply-connected at infinity.

These observations tamed the zoo of open contractible manifolds in some sense. The homotopical characterization of Euclidean space above is sometimes called the open Poincaré conjecture. Stallings' result was later generalized by Siebenmann [130] to yield an essentially homotopical characterization of the Euclidean space amongst all open contractible topological n-manifolds. In conjunction with more recent work on the low-dimensional Poincaré conjecture, i.e., the celebrated work of Perelman [31] and Freedman [102], Siebenmann's theorem can be extended to the following statement.

Theorem 2 *Suppose $n \geq 0$ is an integer.*

1. *For $n \leq 2$, Euclidean space $\mathbb{R}^n$ is the unique open contractible n-manifold.*
2. *For $n \geq 3$, Euclidean space $\mathbb{R}^n$ is the unique open contractible n-manifold that is simply connected at infinity.*
3. *If $n \geq 3$, and M is an open contractible n-manifold, then $M \times \mathbb{R}$ is homeomorphic to $\mathbb{R}^{n+1}$.* $\qquad\qquad\square$

The modern study of open manifolds formalizes the notion of "behavior at ∞" of an open manifold as follows. If W is a locally compact Hausdorff space, we write $\dot{W}$ for a 1-point compactification of W. Write $[0, \infty]$ for the extended real line (i.e., with a limit point at infinity). The end space $e(W)$ consists of the space of maps of pairs $\omega : ([0, \infty], \{\infty\}) \to (\dot{W}, \{\infty\})$ such that $\omega^{-1}(\infty) = \infty$ [76, §1, Definition 1.2]. The end space may be viewed as a homotopy theoretic model for the behaviour at ∞ of W. If W is already compact, then $\dot{W} = W_+$, so the end space is empty. The end space may not be connected: $e(\mathbb{R})$ is homotopy equivalent to S^0 corresponding to the two infinite simple edge paths starting at $0 \in \mathbb{R}$. More generally, $e(\mathbb{R}^n) \simeq S^{n-1}$. In that case, we will say that W is simply connected at infinity if $e(W)$ is connected and simply connected.

5.1.2 Contractible Algebraic Varieties

Cancellation questions like those mentioned above were also studied in algebraic geometry, though in several different contexts. Perhaps the first such question was posed by Zariski [128]: if k is a field, L and L' are finitely generated extensions of k such that the purely transcendental extensions $L(x)$ and $L'(x)$ are isomorphic as extensions of k, are L and L' isomorphic? Phrased in the language of algebraic geometry: if X and X' are irreducible algebraic varieties over a field k such that $X \times \mathbb{P}^1$ is k-birationally equivalent to $X' \times \mathbb{P}^1$, is X k-birational to X'? At some point, in the late 1960s/early 1970s, a closely related "biregular" form of Zariski's original cancellation question was posed: if A and B are finitely generated k-algebras, and $A[x] \cong B[x]$ as k-algebras, is $A \cong B$? Even at the beginning, special attention was paid to the case where A is a polynomial ring. M.P. Murthy [118] also asked: if A is a Krull dimension 2 extension of a field k and $A[x] \cong k[x_1, x_2, x_3]$, then must A be a polynomial ring in 2-variables? In the language of algebraic geometry this leads to the following question.

Question 3 (Biregular Cancellation Question) If X is a smooth affine variety over a field such that $X \times \mathbb{A}^n \cong \mathbb{A}^N$, then is X isomorphic to affine space?   □

Remark 4 The more general cancellation question will be of interest to us as well, and we refer the reader to Sects. 5.3.3 and 5.3.4 for more discussion and history of these kinds of questions. The above question is also sometimes called the "Zariski cancellation problem", but it appears never to have been explicitly posed or considered by Zariski.   □

The situation surrounding Question 3 is quite different depending on the characteristic. Over fields of positive characteristic, recent work of N. Gupta, building on some old ideas of Asanuma [6] exploiting interesting pathologies existing only in positive characteristic produced counterexamples to the biregular cancellation question [59]. Nevertheless, at the time of writing, the question is still open over fields having characteristic 0.

From the outset, ideas of topology played a role in approaches to the biregular cancellation problem over fields having characteristic 0. If $k = \mathbb{C}$, then we write X^{an} for the set $X(\mathbb{C})$ equipped with its usual structure of a complex manifold. More generally, if k is a field for which we can find an embedding $\iota : k \hookrightarrow \mathbb{C}$, then we may use ι to define an associated complex manifold X_ι^{an}. If k is a field that admits a complex embedding, then we will say that X is *topologically contractible* if X_ι^{an} is a contractible space for every complex embedding ι of k. The variety $\mathbb{A}_k^n$ is the primordial example of a topologically contractible variety. Thus, any complex variety that is stably isomorphic to $\mathbb{A}^n$ (see Definition 5.3.3.1) is also automatically topologically contractible. Bearing this in mind, we would like to place the discussion of the biregular cancellation question in context using the theory of open contractible manifolds as a template.

Ramanujam writes in [118] that it was initially hoped that a 2-dimensional non-singular, affine, rational, topologically contractible complex variety would be isomorphic to $\mathbb{A}^2$. However, he constructs a counter-example to this hope in his landmark paper–a surface now called the Ramanujam surface in his honor; we briefly recall the construction of the Ramanujam surface. Ramanujam also gave a topological characterization of the affine plane among complex algebraic varieties: $\mathbb{C}^2$ is the unique non-singular contractible complex surface that is simply connected at infinity.

Example 5 In $\mathbb{P}^2$ take a cubic C with a cusp at q. Let Q be an irreducible conic meeting C with multiplicity 5 at a point p and transversally at a point r. The existence of Q can be deduced from the group law on the non-singular part of C, or directly by avoiding that p is a flex-point of C. Let X be the blow-up of $\mathbb{P}^2$ at r and let $\tilde{C}$, $\tilde{Q}$ denote the strict transforms of C, Q, respectively. The Ramanujam surface is defined by setting

$$\mathcal{R} = X \smallsetminus \tilde{C} \cup \tilde{Q}.$$

Smooth complex varieties X that are not isomorphic to $\mathbb{A}^n$ but for which X^{an} is diffeomorphic to $\mathbb{R}^{2n}$ are known as exotic affine spaces. Such varieties are automatically topologically contractible; for a survey we refer the reader to [156]. After the work of Ramanujam, many examples of topologically contractible smooth varieties were invented; for example, see the work of tom Dieck and Petrie [138] for examples given by explicit equations. In fact, there is a veritable zoo of such examples: in the references just cited, one finds arbitrary dimensional moduli of topologically contractible smooth complex affine surfaces!

The biregular cancellation problem was later solved in the affirmative for surfaces by Miyanishi–Sugie [104] and Fujita [53] in characteristic 0 and Russell in positive characteristic: the affine plane (over an algebraically closed field) is the unique surface that is stably isomorphic to an affine space (see Sect. 5.5.1 for precise references).

Ramanujam's topological characterization of the affine plane also fails to hold in higher dimensions. Indeed, Dimca and Ramanujam both proved that if X is a topologically contractible smooth complex affine variety of dimension $d \geq 3$, then X is automatically diffeomorphic to $\mathbb{R}^{2d}$. Putting these facts together, one sees that there are many exotic affine varieties in dimensions ≥ 3: simply take the products of topologically contractible surfaces and affine spaces! This observation, together with the fact that "topological" invariants do not always have natural counterparts in positive characteristic, suggest that one might hope for a purely algebraic theory of contractible varieties that is more refined than the topological notion discussed here. Nevertheless, there are a great number of questions about the geometry of topologically contractible varieties that one may pose here, and we highlight several questions that have guided the development of the theory of contractible algebraic varieties.

Question 6 (Generalized Serre Question) If X is a topologically contractible smooth complex affine variety, then is every algebraic vector bundle on X trivial?

$\square$

Remark 7 This question is called the generalized Serre question in the literature [156, §8] and we have followed this terminology here. To the best of our knowledge, Serre only ever considered this question for affine space over a field [129, p. 243], where it was spectacularly resolved (independently) by Quillen and Suslin [117, 133]; we refer the reader to [92] for a nice survey of these results. Of course, it is based on the fact that complex topological vector bundles on contractible varieties are trivial. We will see later that the affineness assumption in the question is essential; see Example 5.3.5.4 for more details. $\square$

Question 8 (Generalized van de Ven Question) If X is a topologically contractible smooth complex variety, then is X rational? $\square$

Remark 9 This question is sometimes referred to as the van de Ven Question in the literature [156, Remark 2.2] with reference given to [140]. We call it the generalized van de Ven question because van de Ven was explicitly concerned with surfaces: he writes [140, §3 p.197–198] "All the known examples of algebraic non-singular compactifications of homology 2-cells are rational surfaces. It would follow that there are no others, at least if we restrict ourselves to simply connected homology 2-cells, if the following statement would be true: ($\mathbb{C}$) Every simply connected non-singular algebraic surface with geometric genus 0 is rational." Unfortunately, the conjecture ($\mathbb{C}$) was later disproved: for example the Barlow surfaces [23] are simply connected surfaces of general type with geometric genus 0. Nevertheless, Gurjar and Shastri showed [64, 65] that non-singular contractible surfaces are always rational by a careful case-by-case analysis. $\square$

A theorem of Fujita shows that all topologically contractible smooth complex surfaces are necessarily affine. A remarkable example of Winkelmann [153] shows that Fujita's statement does not hold in higher dimensions. Winkelmann gave the first example of a quasi-affine but not affine smooth contractible complex variety.

Example 10 Winkelmann defines a scheme-theoretically free action of the additive group $\mathbb{G}_a$ on $\mathbb{A}^5$ as follows. Consider the standard 2-dimensional representation V of $\mathbb{G}_a$ given by the embedding $\mathbb{G}_a \hookrightarrow SL_2$ as upper triangular matrices. The 6-dimensional affine space attached to $V \oplus V \oplus V$ thus carries a linear representation of $\mathbb{G}_a$. If we pick coordinates x_0, x_1, y_0, y_1 and z_0, z_1 on this affine space, then x_0, y_0 and z_0 are degree 1 invariant functions, while $x_0 y_1 - y_0 x_1$, $x_0 z_1 - z_0 x_1$ and $y_0 z_1 - z_0 y_1$ are degree 2 invariants. The hypersurface $x_0 = 1 - y_0 z_1 - z_0 y_1$ is thus a $\mathbb{G}_a$-invariant smooth hypersurface in $\mathbb{A}^6$ isomorphic to $\mathbb{A}^5$. The induced action is scheme-theoretically free, and the invariant computation just mentioned identifies the quotient as an open subscheme of a smooth affine quadric of dimension 4 with complement a codimension 2 affine space.   □

Winkelmann's example highlights another important issue in algebraic geometry that is not visible in the topological story. Indeed, it follows from the theorems we stated before that every open contractible manifold may be realized as a quotient of $\mathbb{R}^n$ by a free action of the additive group $\mathbb{R}$. However, every principal $\mathbb{R}$-bundle on a topological space having the homotopy type of a CW-complex is trivial because $\mathbb{R}$ is contractible. In contrast, in algebraic geometry, a natural analog of free $\mathbb{R}$-actions on $\mathbb{R}^n$ is given by (scheme-theoretically) free actions of the additive group scheme $\mathbb{G}_a$ on $\mathbb{A}^n$. One must first ask then: given a scheme-theoretically free action of $\mathbb{G}_a$, when does a quotient even exist as a scheme?[1] If X is a scheme, then the set of isomorphism classes of $\mathbb{G}_a$-torsors on X is parameterized by the sheaf cohomology group $H^1(X, \mathbb{G}_a) = H^1(X, \mathscr{O}_X)$. On an affine scheme, this cohomology group necessarily vanishes, but it need not vanish on a quasi-affine scheme that is not affine.

Example 10 gives the primordial example of a smooth variety that is "algebraically" homotopy equivalent to affine space, without being isomorphic to affine space. While there are various "naive" notions of algebraic homotopy equivalence (e.g., one may think of homotopies parameterized by the affine line), a robust homotopy theory for algebraic varieties in which homotopies are parameterized by the affine line requires more formal machinery. In the sequel, we begin by trying to highlight the key points of one solution–the construction of the Morel–Voevodsky $\mathbb{A}^1$-homotopy [111]–from the standpoint of the tools required by an end user interested in algebro-geometric questions like those posed above. Section 5.2 contains motivation and the basic ideas of the construction, without going into any of the formal categorical preliminaries. In Sect. 5.3, we try to understand concretely and geometrically how to study isomorphisms in the $\mathbb{A}^1$-homotopy category, a.k.a. $\mathbb{A}^1$-weak equivalence. In the context of the discussion above, we introduce the key notion of an $\mathbb{A}^1$-contractible space, see Definition 5.3.1.1. In what remains, we review the theory of $\mathbb{A}^1$-contractible smooth varieties, guided by the discussion above, especially the theory of open-contractible manifolds. Along the way, we discuss an algebro-geometric version of Theorem 1.

Theorem 11 (See Theorems 5.3.5.2 and 5.5.4.3) *Assume k is a field. For every pair of integers $d \geq 3$ and $n \geq 0$, there exists a connected n-dimensional scheme S and a smooth morphism $\pi : X \to S$ of relative dimension d whose fibers are $\mathbb{A}^1$-contractible. Moreover, the fibers over distinct k-points of π are pairwise non-isomorphic.*   □

---

[1] a quotient always exists as an algebraic space.

Much of the rest of the discussion can be viewed as providing background for the following questions, which suggest a best-possible approximation to an algebro-geometric variant of Theorem 2 (see Definition 5.3.3.1 for the terminology).

Question 12 If k is a perfect field, is $\mathbb{A}^n$ the only $\mathbb{A}^1$-contractible smooth k-scheme of dimension ≤ 2?   □

Question 13 Suppose X is an $\mathbb{A}^1$-contractible smooth scheme.

1. Does there exist an affine space $\mathbb{A}^N$ and a Nisnevich locally trivial smooth morphism $\mathbb{A}^N \to X$?
2. If X is moreover affine, then is X a retract of $\mathbb{A}^N$?   □

The first question above is discussed in detail in Sect. 5.5.2. The second is motivated by the discussion of Sect. 5.3.5 in conjunction with the discussion of Sects. 5.3.3, 5.5.3, 5.5.4, and 5.5.5. Finally, we suggest that there is a rather subtle relationship between topologically contractible varieties and $\mathbb{A}^1$-contractible varieties involving "suspension" in the $\mathbb{A}^1$-homotopy category.

Conjecture 14 (See Conjecture 5.5.3.11 and Remark 5.5.3.12) If X is a topologically contractible smooth complex variety with a chosen base-point $x \in X(\mathbb{C})$, then there exists an integer $n \geq 0$ such that $\Sigma^n_{\mathbb{P}^1}(X, x)$ is $\mathbb{A}^1$-contractible; in fact, $n = 2$ should suffice.   □

As with any survey, this one reflects the knowledge and biases of the authors. The literature in affine algebraic geometry on cancellation and related questions is vast, and we can only apologize to those authors whose work we have (inadvertently) failed to appropriately credit.

5.2 A User's Guide to $\mathbb{A}^1$-homotopy Theory

We began by stating a slogan, loosely paraphrasing the first section of [111]:

> there should be a homotopy theory for algebraic varieties over a base where the affine line plays the role assigned to the unit interval in topology.

The path to motivate the construction of the $\mathbb{A}^1$-homotopy category that we follow is loosely based on the work of Dugger [48]. The original constructions of the $\mathbb{A}^1$-homotopy category rely on [80] and are to be found in [111]. General overviews of $\mathbb{A}^1$-homotopy theory may be found in [141], and, especially for "unstable" results in [109]. Morel's paper [107] provides an introductory text, but recent advances like [17] and [68] allow one to get to the heart of the matter much more quickly. We encourage the reader to consult [5, 79, 95], and [151] for recent surveys.

5.2.1   Brief Topological Motivation

The jumping off point for the discussion was the classical Brown-representability theorem in unstable homotopy theory. Recall that the ordinary homotopy category, denoted here $\mathcal{H}$, has as objects "sufficiently nice" topological spaces Top (including, for example, all CW complexes), and morphisms given by homotopy classes of continuous maps between spaces. Unfortunately, the category of CW complexes itself is not "categorically good" enough to build the homotopy category. Indeed, there are various constructions one wants to perform in topology (quotients, loop spaces, mapping spaces) that do not stay in the category of CW complexes (they only stay in the category up to homotopy). We will refer to objects of the category Top (which we have not precisely specified) as *spaces*. As usual, for two spaces X and Y, we write $[X, Y]$ for the set of morphisms between X and Y in the homotopy category.

Suppose $\mathbf{C}$ is a category of "algebraic structures". In practice, one may take $\mathbf{C}$ to be the category of sets or an abelian category like abelian groups, or chain complexes, but we will need some flexibility in the choice. A $\mathbf{C}$-valued invariant is then a contravariant functor $\mathcal{F} \colon \mathrm{Top} \to \mathbf{C}$. We will furthermore consider $\mathbf{C}$-valued invariants on Top that satisfy the following properties:

(i)   (Homotopy invariance axiom) If X is a topological space, and $I = [0, 1]$ is the unit interval, then the map $\mathcal{F}(X) \to \mathcal{F}(X \times I)$ is a bijection.

(ii) (Mayer-Vietoris axiom) If X is a CW complex covered by subcomplexes U and V with intersection $U \cap V$, then we have a diagram of the form

$$\mathcal{F}(X) \to \mathcal{F}(U) \times \mathcal{F}(V) \to \mathcal{F}(U \cap V),$$

and given $u \in \mathcal{F}(U)$ and $v \in \mathcal{F}(V)$, such that the images of u and v under the right hand map coincide, then there is an element of $\mathcal{F}(X)$ whose image under the first map is the pair (u, v).

(iii) (Wedge axiom) The functor $\mathcal{F}$ takes sums to products.

Given a $\mathbf{C}$-valued invariant $\mathcal{F}$ on Top as above, the first condition implies that $\mathcal{F}$ factors through a functor $\mathcal{H} \to \mathbf{C}$, i.e., $\mathcal{F}$ is a $\mathbf{C}$-valued homotopy invariant. There is a very natural class of *representable* Set-valued homotopy invariants, given by $[-, Y]$ for some topological space Y. Given a $\mathbf{C}$-valued homotopy invariant $\mathcal{F}$, the classical Brown representability theorem says that if it satisfies the second and third conditions above, then this functor is a representable homotopy invariant, i.e., there is a CW complex Y and an isomorphism of functors $\mathscr{F} \cong [-, Y]$.

Remark 5.2.1.1 In general, "cohomology theories" satisfy more properties than just those mentioned: e.g., one has a Mayer–Vietoris long exact sequence. For example, consider the functor sending a space to its usual integral singular cochain complex $S^*(X)$. If X is a topological space, then $S^*(X) \to S^*(X \times I)$ is not an isomorphism of chain complexes, but a chain-homotopy equivalence. Likewise, if $X = U \cup V$, then there is a sequence

$$S^*(X) \longrightarrow S^*(U) \oplus S^*(V) \longrightarrow S^*(U \cap V),$$

but this sequence fails to be an exact sequence of chain complexes. To get a Mayer–Vietoris sequence one must replace $S^*(X)$ by a suitable subcomplex with the same cohomology. These observations necessitate making various constructions that are "homotopy invariant" from the very start. $\square$

Remark 5.2.1.2 If a "cohomology theory" is represented on CW complexes by a space Z, and the cohomology theory is geometrically defined (e.g., topological K-theory with Z the infinite grassmannian), then the "representable" cohomology theory extended to all topological spaces need not coincide with the geometric definition for spaces that are not CW complexes. $\square$

5.2.2 Homotopy Functors in Algebraic Geometry

We would like to guess what properties the "homotopy category" will have based on the known invariance properties of cohomology theories in algebraic geometry. To this end, we must have some actual "cohomology theories" at hand. The examples we want to use are the theory of (Bloch's higher) Chow groups [55] and (higher) algebraic K-theory [116], though the reader not familiar with general definitions could focus on the Picard group, which is related to both. We begin by formulating a notion of homotopy invariance in algebraic geometry. As above, suppose **C** is a category of algebraic structures.

Definition 5.2.2.1 A **C**-valued contravariant functor $\mathcal{F}$ on Sm_k is $\mathbb{A}^1$-*invariant* if the morphism $\mathcal{F}(U) \to \mathcal{F}(U \times \mathbb{A}^1)$ is a bijection. $\square$

Example 5.2.2.2 The functor $Pic(X)$ is not $\mathbb{A}^1$-invariant on schemes with singularities that are sufficiently complicated [139]. More generally, Chow groups are $\mathbb{A}^1$-invariant [55, Theorem 3.3] for regular schemes. Likewise, Grothendieck established an $\mathbb{A}^1$-invariance property for algebraic K_0 on regular schemes, and Quillen established a homotopy invariance property for algebraic K-theory of regular schemes (this is one of the fundamental properties of higher algebraic K-theory proven in [116]). $\square$

In the examples above, $\mathbb{A}^1$-invariance, failed to hold on all schemes. As a consequence, we restrict our attention to the category Sm_k of schemes that are separated, smooth and have finite type over k (we use smooth schemes rather than regular schemes since smoothness is more functorially well-behaved than regularity; if we assume we work with varieties over a perfect field, then there is no need to distinguish between the two notions). Our restriction to smooth schemes will be analogous to the restriction to (finite) CW complexes performed above.

We would like to impose some Mayer–Vietoris-like condition on **C**-valued invariants. The most obvious choice would involve the Zariski topology on schemes. In practice, a number of classical algebro-geometric cohomology theories have a Mayer–Vietoris property for a Grothendieck topology that is finer than the Zariski topology. Indeed, Chow groups and algebraic K-theory have "localization" exact sequences; we review such sequences here as a motivation for the finer topology.

Let us recall the localization sequence for Chow groups: if X is a smooth variety, and $U \subset X$ is an open subvariety with closed complement Z (say equi-dimensional of codimension d), there is an exact sequence of the form

$$CH^{*-d}(Z) \longrightarrow CH^*(X) \longrightarrow CH^*(U) \longrightarrow 0;$$

to extend this sequence further to the left, one needs to introduce Bloch's higher Chow groups [27, 28], but we avoid discussing this here. We leave the reader the exercise of showing that from this localization sequence, one may formally deduce that Chow groups and algebraic K-theory have a suitable Mayer–Vietoris property for Zariski open covers by two sets.

One often considers the étale topology in algebraic geometry, and one might ask whether there is an appropriate Mayer-Vietoris sequence for étale covers. In this direction, consider the following situation. Suppose given an open immersion $j : U \hookrightarrow X$ and an étale morphism $\varphi : V \to X$ such that the pair (j, φ) are jointly surjective and such that the induced map $\varphi^{-1}(X \smallsetminus U) \to X \smallsetminus U$ is an isomorphism, diagrammatically this is a picture of the form:

$$
\begin{array}{ccc}
U \times_X V & \xrightarrow{\;j'\;} & V \\
\downarrow{\scriptstyle\varphi} & & \downarrow{\scriptstyle\varphi} \\
U & \xrightarrow{\;j\;} & X
\end{array}
$$

We will refer to such diagrams as *Nisnevich distinguished squares*. One can show that X is the *colimit* in the category of smooth schemes of the diagram $U \longleftarrow U \times_X V \longrightarrow V$.

By a straight-forward diagram chase, one may show that the sequence

$$CH^*(X) \to CH^*(U) \oplus CH^*(V) \to CH^*(U \times_X V)$$

is exact: given an element (u, v) in $CH^*(U) \oplus CH^*(V)$, if the restriction of (u, v) to $CH^*(U \times_X V)$ is zero, then there is an element x in $CH^*(X)$ whose restriction to $CH^*(U) \oplus CH^*(V)$ is (u, v). Therefore, any theory that satisfies localization will have the Mayer–Vietoris property with respect to a Grothendieck topology on schemes that is finer than the Zariski topology.

Example 5.2.2.3 Suppose k is a field of characteristic unequal to 2. Consider the diagram where $X = \mathbb{A}^1$, $U = \mathbb{A}^1 \smallsetminus \{1\}$, $V = \mathbb{A}^1 \smallsetminus \{0, -1\}$. Let j be the usual open immersion of $\mathbb{A}^1 \smallsetminus \{1\}$ into $\mathbb{A}^1$, and let φ be the étale map given by the composite $\mathbb{A}^1 \smallsetminus \{0, -1\} \hookrightarrow \mathbb{G}_m \to \mathbb{G}_m \hookrightarrow \mathbb{A}^1$, where the map $\mathbb{G}_m \to \mathbb{G}_m$ is $z \mapsto z^2$. It is easily checked that this diagram provides a square as above.   □

To define a homotopy category for smooth schemes, informed by the above observations, we will attempt to build a category through which any $\mathbb{A}^1$-homotopy invariant on smooth schemes satisfying Mayer–Vietoris in the Nisnevich sense prescribed above factors. We do this in a few stages. Just as the category of CW-complexes was not "categorically good", we will first enlarge the category of schemes to a suitable category of spaces. Then, we will force Mayer–Vietoris for Nisnevich covers and then impose $\mathbb{A}^1$-homotopy invariance.

Remark 5.2.2.4 As is hopefully evident, we have made a number of choices here: a category of schemes with which to begin, and a topology for which we would like to impose Mayer–Vietoris; we have motivated a particular choice here, but other choices are often warranted. For example, we might want to make constructions involving non-smooth schemes and be able to compare this situation with the one we alluded to above. For this reason, we will try to leave some flexibility in the constructions.   □

5.2.3   The Unstable $\mathbb{A}^1$-homotopy Category: Construction

Spaces

For concreteness, fix a base Noetherian commutative unital ring k of finite Krull dimension. In practice, k will be a field, but for comparing constructions in different characteristics, it will often be useful for k to be a discrete valuation ring or the integers $\mathbb{Z}$. Write Sm_k for the category of schemes that are separated, smooth and have finite type over k. Write Sch_k for the category of Noetherian k-schemes of finite Krull dimension. Write sSet for the category of simplicial sets. There is a functor Set $\to$ sSet sending a set S to the corresponding constant simplicial set (i.e., all face and degeneracy maps are the identity), and we use this functor to identify Set as a full subcategory of sSet.

Definition 5.2.3.1 Write Spc_k for the category of simplicial presheaves on Sm_k, and Spc'_k for the category of simplicial presheaves on Sch_k; objects of these categories will be called *motivic spaces* or k-spaces, depending on whether we want to explicitly specify k.   □

Sending a (smooth) scheme to its corresponding representable presheaf (of constant simplicial sets) defines a functor $\mathrm{Sch}_k \to \mathrm{Spc}'_k$ ($\mathrm{Sm}_k \to \mathrm{Spc}_k$) that is fully-faithful by the Yoneda lemma. We use these functors without mention to identity (smooth) schemes as spaces. Likewise, there is a functor sSet $\to$ Spc_k sending a simplicial set to the corresponding constant simplicial presheaf.

There are many constructions we may perform in the category of spaces: these categories are both complete and cocomplete, i.e., have limits and colimits indexed by all small categories. The category Spc_k (resp. Spc_k') has a final object typically denote $*$ and an inital object $\emptyset$. Using the fact that Spc_k has all small limits and colimits, the following definitions make sense. A pointed space is a pair $(\mathcal{X}, x)$ where $\mathcal{X} \in \mathrm{Spc}_k$ and $x : * \to \mathcal{X}$ is a morphism of spaces. We write $\mathrm{Spc}_{k,\bullet}$ for the category of *pointed k-spaces*. It is important to emphasize that these constructions are being made in the category of *spaces* and NOT in the category of schemes. Moreover, moving outside of the category of schemes has a number of tangible benefits.

- If $(\mathcal{X}, x)$ and $(\mathcal{Y}, y)$ are pointed k-spaces, then the wedge sum $\mathcal{X} \vee \mathcal{Y}$ is the pushout (colimit) of the diagram $\mathcal{X} \xleftarrow{x} * \xrightarrow{y} \mathcal{Y}$.
- If $(\mathcal{X}, x)$ and $(\mathcal{Y}, y)$ are pointed k-spaces, then the smash product $\mathcal{X} \wedge \mathcal{Y}$ is the quotient of $\mathcal{X} \times \mathcal{Y} / \mathcal{X} \vee \mathcal{Y}$.
- We write S^i for the constant presheaf with value $\Delta^i / \partial \Delta^i$, where Δ^i is the usual i-simplex.

Nisnevich and cdh Distinguished Squares

We now make precise various ideas given above motivated by Mayer–Vietoris sequences.

Definition 5.2.3.2 A Nisnevich distinguished square is a pull-back diagram of schemes

$$
\begin{array}{ccc}
W & \xrightarrow{j'} & V \\
\downarrow{\varphi} & & \downarrow{\varphi} \\
U & \xrightarrow{j} & X
\end{array}
$$

such that $j : U \hookrightarrow X$ is an open immersion, $\varphi : V \to X$ is an étale morphism, and the induced map $\varphi^{-1}(X \smallsetminus U) \to X \smallsetminus U$ is an isomorphism of schemes given the reduced induced scheme structure. □

Remark 5.2.3.3 Every Zariski open cover of a scheme by two open sets gives rise to a Nisnevich distinguished square. Nisnevich distinguished squares generate a Grothendieck topology on Sm_k or Sch_k called the Nisnevich topology. This topology is conveniently described in terms of covering sieves: it is the coarsest topology such that the empty sieve covers $\emptyset$, and for every distinguished square as above, the sieve on X generated by $U \to X$ and $V \to X$ is a covering sieve. This definition of the Nisnevich topology is equivalent to other standard ones in the literature. In fact, the Nisnevich topology may be generated by much simpler squares where all corners of the square are affine and the reduced closed complement of U in X is given by a principal ideal. We refer the interested reader to [17, §2] for more details.   □

For the most part, we will only use Nisnevich distinguished squares and the Nisnevich topology, but the following definition will also be useful.

Definition 5.2.3.4 An abstract blow-up square is a pull-back diagram of schemes of the form:

$$
\begin{array}{ccc}
E & \longrightarrow & Y \\
\downarrow & & \downarrow{\scriptstyle \pi} \\
Z & \overset{i}{\longrightarrow} & X
\end{array}
$$

where i is a closed immersion, π is a proper map, and the induced map $\pi^{-1}(X \smallsetminus i(Z)) \to X \smallsetminus i(Z)$ is an isomorphism.   □

Remark 5.2.3.5 As above, the proper cdh topology on Sch_k is the coarsest topology such that the empty sieve covers $\emptyset$, and for which the sieve on X generated by $Z \to X$ and $Y \to X$ is a covering sieve. The cdh topology is the smallest Grothendieck topology whose covering morphisms include those of the proper cdh topology and those of the Nisnevich topology.   □

Localization

Given the category Spc_k (resp. Spc'_k), we now want to formally invert a set of morphisms $\mathscr{S}$ to build a homotopy category. Each covering sieve of a Grothendieck topology gives rise to a monomorphism of presheaves with target a representable presheaf; to impose "Mayer–Vietoris" with respect to a given topology, one first formally inverts the set of all of these monomorphisms. Concretely, if $u : U \to X$ is a Nisnevich covering, then one may build a simplicial presheaf $\check{C}(U)$ by taking iterated fiber products of U with itself over X. The morphism u then yields an augmentation

$$\check{C}(U) \longrightarrow X$$

that we would like to force to be a weak equivalence. The Nisnevich (resp. cdh) local homotopy category is, in essence, the universal category where the above maps have been inverted (see [17, Lemma 3.1.3] for a precise statement). There are numerous constructions of this category now and the "universal" point-of-view espoused here was studied in great detail by D. Dugger; see [48] for more details.

One standard way to invert the relevant set of morphisms is to equip Spc_k (resp. Spc'_k) with the structure of model category (see, e.g., [72]); this involves specifying classes of cofibrations and fibrations along with the classes of morphisms that are to be inverted, i.e., the morphisms we want to become weak equivalences. In any case, we will write $H_{Nis}(k)$ (resp. $H_{cdh}(k)$) for the corresponding local homotopy category; we refer the reader to [83] for a textbook treatment of local homotopy theory. By construction, there is a functor $Spc_k \rightarrow H_{Nis}(k)$ (resp. $Spc'_k \rightarrow H_{cdh}(k)$) that is suitably (i.e., homotopically) initial.

One may build $\mathbb{A}^1$-homotopy categories in a similar universal fashion. After formally inverting Nisnevich (resp. cdh) local weak equivalences, we further invert the projection from the affine line:

$$\mathscr{X} \times \mathbb{A}^1 \longrightarrow \mathscr{X}.$$

We write $H_{\mathbb{A}^1}(k)$ for the corresponding category obtained by localizing $H_{Nis}(k)$ and $H_{\mathbb{A}^1}^{cdh}(k)$ for the category obtained by localizing $H_{cdh}(k)$.

Notation 5.2.3.6 An isomorphism in $H_{\mathbb{A}^1}(k)$ or $H_{\mathbb{A}^1}^{cdh}(k)$ will be called an $\mathbb{A}^1$-weak equivalence. To emphasize the analogy with topology, we write $[\mathscr{X}, \mathscr{Y}]_{\mathbb{A}^1}$ for the set of morphisms in either category; we will read this as the set of $\mathbb{A}^1$-homotopy classes of maps between $\mathscr{X}$ and $\mathscr{Y}$.   □

Remark 5.2.3.7 Given $\mathscr{X}$ and $\mathscr{Y}$ in Spc_k and a morphism $H : \mathscr{X} \times \mathbb{A}^1 \rightarrow \mathscr{Y}$, we may think of H as an algebraic homotopy between $H(-, 0) : \mathscr{X} \rightarrow \mathscr{Y}$ and $H(-, 1) : \mathscr{X} \rightarrow \mathscr{Y}$; we will say that $H(-, 0)$ and $H(-, 1)$ are naively homotopic. More generally, we will say that two morphisms $\mathscr{X} \rightarrow \mathscr{Y}$ are *naively homotopic* if they are equivalent for the equivalence relation generated by naive homotopies. Note that naively homotopic maps determine the same $\mathbb{A}^1$-homotopy class, i.e., there is an evident function

$$Hom_{Spc_k}(\mathscr{X}, \mathscr{Y})/\{naive \ \mathbb{A}^1\text{-homotopies}\} \longrightarrow [\mathscr{X}, \mathscr{Y}]_{\mathbb{A}^1}.$$

Because of the relatively abstract definition of the target of this morphism, it is an interesting question to understand situations in which this morphism is a bijection.   □

5.2.4 The Unstable $\mathbb{A}^1$-homotopy Category: Basic Properties

Motivic Spheres

In addition to the simplicial circle S^1 described above, we introduce the Tate circle $\mathbb{G}_m$, viewed as a space pointed by the identity section. We then define bigraded motivic spheres by the formula

$$S^{i,j} = S^i \wedge \mathbb{G}_m^{\wedge j}.$$

From these definitions, and basic homotopy-invariance statements, one may write down some standard algebro-geometric models of motivic spheres. Homotopy categories typically do not have robust notions of limit or colimit, and one therefore considers a more flexible "up to homotopy" notion. The results below are straightforward homotopy colimit computations, and we mention the required facts without going into details.

Example 5.2.4.1 The space $\mathbb{A}^1/\{0, 1\}$ is a model for S^1. This is an algebro-geometric analog of the fact that the circle S^1 can be realized as the homotopy pushout of the two-point space S^0 along the maps projecting onto each point.   □

One of the basic consequences of our construction of the $\mathbb{A}^1$-homotopy category is that any Zariski Mayer–Vietoris square is a pushout square.

Example 5.2.4.2 There is an $\mathbb{A}^1$-weak equivalence $S^{1,1} \cong \mathbb{P}^1$. To see this, use the standard Zariski open cover of $\mathbb{P}^1$ by two copies of $\mathbb{A}^1$.   □

Example 5.2.4.3 There is an $\mathbb{A}^1$-weak equivalence $\mathbb{A}^n \smallsetminus 0 \cong S^{n-1,n}$. Use induction and the open cover of $\mathbb{A}^n \smallsetminus 0$ by the open sets $(\mathbb{A}^{n-1} \smallsetminus 0) \times \mathbb{A}^1$ and $\mathbb{G}_m \times \mathbb{A}^{n-1}$ with intersection $\mathbb{G}_m \times (\mathbb{A}^{n-1} \smallsetminus 0)$.   □

Example 5.2.4.4 There is an $\mathbb{A}^1$-weak equivalence $\mathbb{P}^n/\mathbb{P}^{n-1} \cong S^{n,n}$. Use the open cover of $\mathbb{P}^n$ by $\mathbb{A}^n$ and $\mathbb{P}^n \smallsetminus 0$ with intersection $\mathbb{A}^n \smallsetminus 0$.   □

Remark 5.2.4.5 With reference to Definition 5.4.1.3 we remark that $S^{i,j}$ is $\mathbb{A}^1$-$(i-1)$-connected in the sense that the $\mathbb{A}^1$-homotopy sheaf $\pi_n^{\mathbb{A}^1}(S^{i,j})$ vanishes for $n \leq i - 1$, see [109, §3]. This corresponds precisely to the connectivity of the real points of $S^{i,j}$ or equivalently the connectivity of the simplicial sphere S^i.   □

Representability Statements

Just like in topology, there is a Brown representability theorem characterizing homotopy functors in algebraic geometry. In addition to homotopy invariance, one wants functors that turn Nisnevich distinguished squares into "homotopy" fiber products; for more details, see the works of Jardine [82] and Naumann-Spitzweck [113]. One additional subtlety that arises is that many natural functors of algebro-geometric origin that arise fail to be $\mathbb{A}^1$-invariant on the category of all smooth schemes. Thus, one would also like to investigate representability questions for such functors. We summarize some theorems that will be useful in the sequel.

In topology, complex line bundles are represented by homotopy classes of maps to infinite complex projective space, at least on spaces having the homotopy type of a CW complex. There is an algebro-geometric analog of this result. For any base commutative unital ring k, one may define the space $\mathbb{P}_k^\infty$ as the colimit of $\mathbb{P}_k^n$ along the standard closed immersions $\mathbb{P}^n \hookrightarrow \mathbb{P}^{n+1}$. The space $\mathbb{P}^\infty$ can be given a multiplication in the $\mathbb{A}^1$-homotopy category. Indeed, the Segre embeddings $\mathbb{P}^n \times \mathbb{P}^m \longrightarrow \mathbb{P}^{(n+1)(m+1)-1}$ may be used to define a map $\mathbb{P}^\infty \times \mathbb{P}^\infty \to \mathbb{P}^\infty$.

This multiplication map may be shown to be associative up to $\mathbb{A}^1$-homotopy and equips $[-, \mathbb{P}^\infty]_{\mathbb{A}^1}$ with a group structure. In the algebro-geometric setting, one has a representability theorem for algebraic line bundles.

Proposition 5.2.4.6 ([111, §4 Proposition 3.8]) *If k is a regular ring, then for any smooth k-scheme X, there is an isomorphism of the form*

$$Pic(X) \xrightarrow{\sim} [X, \mathbb{P}^\infty]_{\mathbb{A}^1};$$

this isomorphism is functorial in X.   □

In topology, complex vector bundles of a given rank are represented by homotopy classes of maps to a suitable infinite Grassmannian, at least on spaces having the homotopy type of a CW complex. Unfortunately, the functor assigning to a smooth scheme X over a base k the set $\mathscr{V}_r(X)$ of isomorphism classes of rank r vector bundles on X fails to be $\mathbb{A}^1$-invariant, as the following example shows. Thus, this functor cannot be representable on the category of *all* smooth schemes.

Example 5.2.4.7 Consider $\mathbb{P}^1$. A classical result of Dedekind–Weber often attributed to Grothendieck asserts that all vector bundles on $\mathbb{P}^1$ are direct sums of line bundles (see [67] for an elementary proof). Consider $X = \mathbb{P}^1 \times \mathbb{A}^1$ with coordinates t and x. The matrix

$$\begin{pmatrix} t^{-1} & x \\ 0 & t \end{pmatrix}$$

is the clutching function for a rank 2 vector bundle on $\mathbb{P}^1 \times \mathbb{A}^1$. The fiber over $x = 0$ of this bundle is isomorphic to $\mathscr{O}(-1) \oplus \mathscr{O}(1)$, while the fiber over $x = 1$ (or any other non-zero value) is isomorphic to $\mathscr{O} \oplus \mathscr{O}$. This rank 2 vector bundle is therefore evidently not pulled back from a vector bundle on $\mathbb{P}^1$.   □

A classical result of Lindel [96] shows that the functor $\mathscr{V}_r(X)$ is $\mathbb{A}^1$-invariant upon restriction to the category of smooth affine k-schemes, if k is a field. This result was extended by Popescu to the case where k is a Dedekind domain with perfect residue fields. Then, even though $\mathscr{V}_r(X)$ fails to be representable on all smooth schemes, one can hope it is representable on smooth affine schemes. On smooth affine schemes, given a vector bundle, one may always choose generating sections: every vector bundle on a smooth affine scheme thus determines (non-uniquely) a morphism to a Grassmannian. We may define Gr_n as a space to be the colimit of $Gr_{n,N}$ over standard inclusions $Gr_{n,N} \to Gr_{n,N+1}$. The morphisms to Gr_n defined by different choices of generating sections all lie in the same $\mathbb{A}^1$-homotopy class. Thus, one obtains a well-defined bijection in the following result.

Theorem 5.2.4.8 (Morel, Schlichting, Asok–Hoyois–Wendt) *Assume k is smooth over a Dedekind ring with perfect residue fields. For any smooth affine k-scheme X,*

the assignment sending a rank r vector bundle on X to the morphism $X \to Gr_r$
corresponding to a choice of generating sections factors through a pointed bijection

$$\mathscr{V}_r(X) \xrightarrow{\sim} [X, Gr_r]_{\mathbb{A}^1};$$

this bijection is functorial in X. □

Proof This result was stated originally by F. Morel [110, Theorem 8.1] in the case where k is perfect and assuming $r \neq 2$. Morel's argument was greatly simplified by M. Schlichting [126, Theorem 6.22]. The result appears in the form above in [17, Theorem 1]. □

Remark 5.2.4.9 While vector bundles of a given rank are only representable on smooth affine schemes, upon passing to stable isomorphism classes, i.e., algebraic K-theory, one may obtain a representability statement on all smooth schemes. Representability of algebraic K-theory was first established in [111, Theorem 3.13], though we refer the reader to [127, Remark 2 p. 1162] for some corrections to the original argument. □

Representability of Chow Groups

Chow cohomology groups are also representable on smooth schemes, but the representing object is a bit more subtle. To set the stage recall that ordinary singular cohomology with coefficients in an abelian group A is representable on finite CW-complexes by Eilenberg–Mac Lane spaces $K(A, n)$. A classical result of Dold and Thom gives a concrete geometric model for the Eilenberg–Mac Lane space $K(\mathbb{Z}, n)$. Indeed for a pointed topological space T, we may define the symmetric product $Sym^n(T)$ as the quotient of the n-fold product by the action of the symmetric group on n letters permuting the factors, i.e.,

$$Sym^n(T) = T^n/\Sigma_n.$$

Using the base-point, there are natural inclusions $Sym^n T \to Sym^{n+1} T$, and one defines the infinite symmetric product $Sym(T)$ as the colimit of the finite symmetric powers with respect to these inclusions. The space $Sym(T)$ may be thought of as the free commutative monoid on T. By "group completing," we may define a space $Sym(T)^+$ that is a topological abelian group. For connected spaces, the group completion process does not alter the homotopy type, and the classical Dold–Thom theorem shows that for every $n \geq 0$, there are weak equivalences of the form:

$$K(\mathbb{Z}, n) \cong Sym(S^n),$$

i.e., Eilenberg–Mac Lane spaces may be realized as infinite symmetric products of spheres.

The procedure sketched above yields a reasonable representing model for Chow groups as well. For a smooth scheme X, define a presheaf $\mathbb{Z}_{tr}(X)$ on Sm_k by assigning to $U \in Sm_k$ the free abelian group on irreducible closed subschemes of $U \times X$ that are finite and surjective over a component of U. This construction is covariantly functorial in X as well, and for a closed subscheme $Z \subset X$, we define $\mathbb{Z}_{tr}(X/Z) = \mathbb{Z}_{tr}(X)/\mathbb{Z}_{tr}(Z)$, where the latter quotient is the quotient as presheaves of abelian groups. In particular, we saw earlier that $\mathbb{P}^n/\mathbb{P}^{n-1}$ is a model of the motivic sphere $S^{n,n}$, and we set

$$K(\mathbb{Z}(n), 2n) := \mathbb{Z}_{tr}(\mathbb{P}^n/\mathbb{P}^{n-1}).$$

The spaces $K(\mathbb{Z}(n), 2n)$ are usually called *motivic Eilenberg–Mac Lane spaces*. With that definition in mind, we can formulate the appropriate representability theorem.

Theorem 5.2.4.10 ([38]) *Assume k is a perfect field. For every $n \geq 0$, and every smooth k-scheme X, there is a canonical bijection*

$$[X_+, K(\mathbb{Z}(n), 2n)]_{\mathbb{A}^1} \longrightarrow CH^n(X).$$

Remark 5.2.4.11 We refer the reader to [144, §2] for a convenient summary of properties of motivic cohomology phrased in terms of motivic Eilenberg–Mac Lane spaces. The relationship between $K(\mathbb{Z}(n), 2n)$ and symmetric products, which may appear a bit obscure above stems from the link between symmetric powers and algebraic cycles on quasi-projective varieties; we refer the interested reader to [134], [141, §6.1] and the references therein for more details. This relationship is perhaps most clearly seen in the case $n = 1$ where $CH^1(X) = Pic(X)$. One may define Sym^n on the category of (say) quasi-projective varieties over a field. It is well-known that $\mathbb{P}^n \cong Sym^n(\mathbb{P}^1)$ as schemes (essentially, this is the map sending a polynomial in 1 variable to its roots). Thus, $Sym(\mathbb{P}^1) \cong \mathbb{P}^\infty$. Using this identification, one may build a map $\mathbb{P}^\infty \to K(\mathbb{Z}(1), 2)$; this map is an $\mathbb{A}^1$-weak equivalence. □

The Purity Isomorphism

Definition 5.2.4.12 If X is a smooth scheme and $\pi : E \to X$ is a vector bundle with zero section $i : X \to E$, we define $Th(\pi) = E/E - i(X)$. □

This definition of Thom space has many of the same properties as the corresponding construction in topology; we summarize some here.

Proposition 5.2.4.13 *Suppose X is a smooth scheme.*

1. If $\pi : X \times \mathbb{A}^n \to X$ is a trivial bundle of rank n, then $Th(\pi) \cong \mathbb{P}^{1 \wedge n} \wedge X_+$.

2. If $\psi : X' \to X$ is a morphism of schemes, $\pi' : E' \to X'$ is a vector bundle on X' and $\pi : E \to X$ is a vector bundle on X fitting into a commutative square of the form

$$
\begin{array}{ccc}
E' & \xrightarrow{\varphi} & E \\
\downarrow{\scriptstyle \pi'} & & \downarrow{\scriptstyle \pi} \\
X' & \xrightarrow{\psi} & X,
\end{array}
$$

where φ is a fiberwise monomorphism, then there is an induced morphism $Th(\pi') \to Th(\pi)$.

3. If $\pi : E \to X$ and $\pi' : E' \to X$ are vector bundles over X, then $Th(\pi \oplus \pi') \cong Th(\pi) \wedge Th(\pi')$. $\qquad\square$

The importance of the notion of Thom space stems from the purity isomorphism, which we summarize in the next result. For later use, we will need to understand functoriality of purity isomorphism. In order to precisely formulate the functoriality properties, we need to introduce a bit of terminology. Suppose we have a cartesian square of smooth schemes of the form

$$
\begin{array}{ccc}
Z' & \xrightarrow{i'} & X' \\
\downarrow & & \downarrow{\scriptstyle f} \\
Z & \xrightarrow{i} & X
\end{array}
$$

where i is a closed immersion. In that case, i' is also a closed immersion and we will say the square is *transversal* if the induced map of normal bundles $\varphi : \nu_{Z'/X'} \to f^* \nu_{Z/X}$ is an isomorphism. The following result was established in [111, §3 Theorem 2.23]; the subsequent functoriality statements appear in [145, §2].

Theorem 5.2.4.14 (Homotopy Purity) *Suppose k is a Noetherian ring of finite Krull dimension.*

1. *If $i : Z \to X$ is a closed immersion of smooth k-schemes, then there is a purity isomorphism*

$$
X/X - i(Z) \cong Th(\nu_i).
$$

2. *If $i : Z \to X$ is the zero section of a geometric vector bundle $\pi : X \to Z$, then the purity isomorphism is the identity map.*

3. *Given a transversal diagram of smooth schemes as above, the purity isomorphism is functorial in the sense that there is a commutative square of the form*

$$
\begin{array}{ccc}
X'/(X' \smallsetminus Z') & \longrightarrow & Th(\nu_{Z'/X'}) \\
\downarrow & & \downarrow \\
X/(X \smallsetminus Z) & \longrightarrow & Th(\nu_{Z/X}),
\end{array}
$$

where the horizontal maps are the purity isomorphisms, the left vertical map is the map on quotients induced by commutativity of the square and the right vertical map is the map induced by functoriality of Thom spaces.   □

Remark 5.2.4.15 As will become clear in the sequel, homotopy purity is an absolutely fundamental tool in $\mathbb{A}^1$-homotopy theory, especially from the standpoint of computations. In the context of stable categories to be introduced in Sect. 5.2.5 it will lead to Gysin exact sequences.   □

Comparison of Nisnevich and cdh-local $\mathbb{A}^1$-weak Equivalences

There is an obvious inclusion $\mathrm{Sm}_k \to \mathrm{Sch}_k$; this yields a functor $\mathrm{Spc}'_k \to \mathrm{Spc}_k$ by restriction. One may show that there is an induced (derived) "pullback" functor $\pi^* : \mathrm{H}_{\mathbb{A}^1}(k) \to \mathrm{H}^{cdh}_{\mathbb{A}^1}(k)$. For later use, we will need the following comparison theorem of Voevodsky [26, Theorem 5.1] [146, Theorem 4.2].

Theorem 5.2.4.16 (Voevodsky) *Assume k is a field having characteristic 0. Suppose $f : (\mathscr{X}, x) \to (\mathscr{Y}, y)$ is a pointed morphism in Spc_k. If $\pi^*(f)$ is an isomorphism in $\mathrm{H}^{cdh}_{\mathbb{A}^1}(k)$, then Σf is an isomorphism in $\mathrm{H}_{\mathbb{A}^1}(k)$.*   □

5.2.5 A Snapshot of the Stable Motivic Homotopy Category

One of the basic lessons of classical homotopy theory is that calculations become more accessible after inverting the suspension functor on the homotopy category of pointed spaces. The notion of a topological spectrum makes this process precise [3]. Similar constructions turn out to be extremely useful in the setting of motivic homotopy theory following [50, 73, 81, 107, 141].

For the purposes of this survey it is useful to know there exists a closed symmetric monoidal category $\mathscr{SH}(k)$ called the stable motivic homotopy category of the field k. We note that $\mathscr{SH}(k)$ is an additive category, in fact a triangulated category equipped with the auto-equivalence given by smashing with the simplicial circle. This category is obtained by formally inverting suspension with the projective line $\mathbb{P}^1$ on the category $\mathrm{Spc}_{k,\bullet}$ of pointed motivic spaces. One formally inverts $\mathbb{P}^1$ by considering "spectra". A motivic spectrum $E \in \mathrm{Spt}_k$ is comprised of pointed motivic spaces E_n for all $n \geq 0$ together with structure maps $\mathbb{P}^1 \wedge E_n \to E_{n+1}$. For example, every $X \in \text{Sm}_k$ has an associated motivic suspension spectrum $\Sigma_{\mathbb{P}^1}^{\infty} X_+$ with terms $(\mathbb{P}^1)^{\wedge n} \wedge X_+$ and identity structure maps. Reminiscent of the way the natural numbers give rise of the integers we are entitled to motivic spheres $S^{p,q}$ in $\mathscr{SH}(k)$ for all $p, q \in \mathbb{Z}$. In fact, $\mathscr{SH}(k)$ is generated by shifted motivic suspension spectra of the form $\Sigma^{p,q} \Sigma_{\mathbb{P}^1}^{\infty} X_+$. With a great deal of tenacity one can make precise the statement that $\mathscr{SH}(k)$ is the associated stable homotopy category of Spt_k. Moreover, one can define a symmetric monoidal structure on $\mathscr{SH}(k)$ for which the sphere spectrum $\mathbf{1} = \Sigma_{\mathbb{P}^1}^{\infty} \text{Spec}(k)_+$ is the unit.

There are standard Quillen adjunctions, whose left adjoints preserve weak equivalences

$$\Sigma_{\mathbb{P}^1} : \text{Spc}_{k,\bullet} \rightleftarrows \text{Spc}_{k,\bullet} : \Omega_{\mathbb{P}^1}$$

$$\Sigma_{\mathbb{P}^1}^{\infty} : \text{Spc}_{k,\bullet} \rightleftarrows \text{Spt}_k : \Omega_{\mathbb{P}^1}^{\infty}.$$

Moreover, we let $\mathcal{H}om(E, F)$ denote the internal homomorphism object of $\mathscr{SH}(k)$ characterized by the adjunction isomorphism

$$\text{Hom}_{\mathscr{SH}(k)}(D, \mathcal{H}om(E, F)) \simeq \text{Hom}_{\mathscr{SH}(k)}(D \wedge E, F).$$

Later on in our discussion of $\mathbb{A}^1$-contractibility we will appeal to the following result connecting the stable and unstable worlds of motivic homotopy theory. We include a proof since it illustrates some basic concepts and techniques.

Lemma 5.2.5.1 *Let X be a smooth scheme and $x \in X$ a closed point. If $\Sigma_{\mathbb{P}^1}^{\infty}(X, x) \simeq *$ in $\mathscr{SH}(k)$, then there exists an integer $n \geq 0$ such that $\Sigma_{\mathbb{P}^1}^n(X, x)$ is $\mathbb{A}^1$-contractible.* □

Proof By [50, Definition 2.10], an object $F \in \text{Spc}_{k,\bullet}$ is fibrant exactly when for every $X \in \text{Sm}_k$, (1) $F(X)$ is a Kan complex; (2) the projection $X \times \mathbb{A}^1 \to X$ induces a homotopy equivalence $F(X) \simeq F(X \times \mathbb{A}^1)$; (3) F maps Nisnevich elementary distinguished squares in Sm_k to homotopy pullback squares of simplicial sets, and $F(\emptyset)$ is contractible. Moreover, a motivic spectrum $E \in \text{Spt}_k$ is fibrant if and only if it is levelwise fibrant and an $\Omega_{\mathbb{P}^1}$-spectrum.

Let $(E_n)_{n \geq 0}$ be a levelwise fibrant replacement of $\Sigma_{\mathbb{P}^1}^{\infty}(X, x)$, i.e., E_n is a fibrant replacement of $\Sigma_{\mathbb{P}^1}^n(X, x)$ in $\text{Spc}_{k,\bullet}$, and let E be a fibrant replacement of $\Sigma_{\mathbb{P}^1}^{\infty}(X, x)$ in Spt_k. A key observation is that filtered colimits in $\text{Spc}_{k,\bullet}$ preserve fibrant objects; this follows from the above description of fibrant objects and the facts that filtered colimits of simplicial sets preserve Kan complexes, homotopy equivalences, and homotopy pullback squares [50, Corollary 2.16]. Putting these facts together, one deduces that there is a simplicial homotopy equivalence

$$\Omega_{\mathbb{P}^1}^{\infty} E \simeq \text{colim}_{n \to \infty} \Omega_{\mathbb{P}^1}^n E_n.$$

Let $\tilde{X} \in \mathrm{Spc}_{k,\bullet}$ be the simplicial presheaf $(X, x) \vee \Delta^1$ pointed at the free endpoint of Δ^1; this is a cofibrant replacement of (X, x) in $\mathrm{Spc}_\bullet(k)$. Since $\tilde{X} \in \mathrm{Spc}_{k,\bullet}$ is finitely presentable, the following are homotopy equivalences of Kan complexes, where Map denotes the simplicial sets of maps in the above simplicial model categories

$$\mathrm{Map}(\Sigma^\infty_{\mathbb{P}^1} \tilde{X}, E) \simeq \mathrm{Map}(\tilde{X}, \Omega^\infty_{\mathbb{P}^1} E)$$

$$\simeq \mathrm{Map}(\tilde{X}, \mathrm{colim}_{n \to \infty} \Omega^n_{\mathbb{P}^1} E_n)$$

$$\simeq \mathrm{colim}_{n \to \infty} \mathrm{Map}(\tilde{X}, \Omega^n_{\mathbb{P}^1} E_n)$$

$$\simeq \mathrm{colim}_{n \to \infty} \mathrm{Map}(\Sigma^n_{\mathbb{P}^1} \tilde{X}, E_n).$$

The hypothesis that $\Sigma^\infty_{\mathbb{P}^1}(X, x)$ is weakly contractible means that the weak equivalence $\Sigma^\infty_{\mathbb{P}^1} \tilde{X} \overset{\sim}{\to} E$ and the zero map $\Sigma^\infty_{\mathbb{P}^1} \tilde{X} \to * \to E$ are in the same connected component of the Kan complex $\mathrm{Map}(\Sigma^\infty_{\mathbb{P}^1} \tilde{X}, E)$. Since π_0 preserves filtered colimits of simplicial sets, there exists an integer $n \geq 0$ such that the weak equivalence $\Sigma^n_{\mathbb{P}^1} \tilde{X} \overset{\sim}{\to} E_n$ and the zero map $\Sigma^n_{\mathbb{P}^1} \tilde{X} \to * \to E_n$ belong to the same connected component of $\mathrm{Map}(\Sigma^n_{\mathbb{P}^1} \tilde{X}, E_n)$. In other words, $\Sigma^n_{\mathbb{P}^1} \tilde{X} \simeq \Sigma^n_{\mathbb{P}^1}(X, x)$ is $\mathbb{A}^1$-contractible. $\square$

Stable Representablity of Algebraic K-theory

Algebraic K-theory is also representable in the stable $\mathbb{A}^1$-homotopy category. To see this, it suffices to consider the infinite projective space $\mathbb{P}^\infty$ and a certain "Bott element" β obtained from the virtual vector bundle $[\mathcal{O}] - [\mathcal{O}(-1)]$ over $\mathbb{P}^1$. The precise context involves the stable motivic homotopy category $\mathscr{SH}(k)$; we replace $\mathbb{P}^\infty$ with its motivic suspension spectrum $\Sigma^\infty_{\mathbb{P}^1} \mathbb{P}^\infty_+$ upon which it makes sense to invert β.

Theorem 5.2.5.2 (Gepner–Snaith, Spitzweck–Østvær) *If k is a noetherian ring with finite Krull dimension, then there is a natural isomorphism in $\mathscr{SH}(k)$*

$$\Sigma^\infty_{\mathbb{P}^1} \mathbb{P}^\infty_+[\beta^{-1}] \cong \mathbf{KGL}.$$

Proof (Comments on the Proof) Here **KGL** is the algebraic K-theory spectrum introduced by Voevodsky [141] (over regular base schemes it represents Quillen's algebraic K-theory). Independent proofs of this result are given in [56, 131]. The conclusion holds more generally over any noetherian base scheme of finite Krull dimension. $\square$

Milnor–Witt K-theory

For later reference we recall the definition of Milnor-Witt K-theory $\mathbf{K}_*^{MW}(k)$ in [110]. It is the quotient of the free associative integrally graded ring on the set of symbols $[k^\times] := \{[u] \mid u \in k^\times\}$ in degree 1 and η in degree -1 by the homogeneous ideal imposing the relations

(1) $[uv] = [u] + [v] + \eta[u][v]$ (η-twisted logarithm),
(2) $[u][v] = 0$ for $u + v = 1$ (Steinberg relation),
(3) $[u]\eta = \eta[u]$ (commutativity), and
(4) $(2 + [-1]\eta)\eta = 0$ (hyperbolic relation).

Milnor-Witt K-theory is ε-commutative for $\varepsilon = -(1 + [-1]\eta)$. By work of Morel there is an isomorphism with the graded ring of endomorphisms of the sphere

$$\mathbf{K}_*^{MW}(k) \cong \bigoplus_{n \in \mathbb{Z}} \pi_{n,n}\mathbf{1}.$$

Moreover, we have $\mathbf{K}_0^{MW}(k) \cong GW(k)$, the Grothendieck-Witt ring of stable isomorphism classes of symmetric bilinear forms [103]. Inverting η in $\mathbf{K}_*^{MW}(k)$ yields the ring of Laurent polynomials $W(k)[\eta^{\pm 1}]$ over the Witt ring, and $\mathbf{K}_*^{MW}(k)/\eta \cong \mathbf{K}_*^M(k)$, the Milnor K-theory ring of k [101].

5.3 Concrete $\mathbb{A}^1$-weak Equivalences

In this section, we attempt to make the discussion of the previous section more concrete. In particular, we will discuss fundamental examples of isomorphisms in the unstable $\mathbb{A}^1$-homotopy category. Moreover, we recall some results from affine (and quasi-affine) algebraic geometry in the context of $\mathbb{A}^1$-homotopy theory.

5.3.1 Constructing $\mathbb{A}^1$-weak Equivalences of Smooth Schemes

By construction of the $\mathbb{A}^1$-homotopy category, for any smooth scheme X the projection map $X \times \mathbb{A}^1 \to X$ is an $\mathbb{A}^1$-weak equivalence; in particular, the morphism $\mathbb{A}^1 \to \operatorname{Spec} k$ is an $\mathbb{A}^1$-weak equivalence. By induction, one concludes that $\mathbb{A}^n \to \operatorname{Spec} k$ is an $\mathbb{A}^1$-weak equivalence. In fact, one may give a completely algebraic construction of this $\mathbb{A}^1$-weak equivalence using the ideas of Remark 5.2.3.7. Indeed, there is a morphism $\mathbb{A}^1 \times \mathbb{A}^n \to \mathbb{A}^n$ sending $(t, x_1, \ldots, x_n) \mapsto (tx_1, \ldots, tx_n)$; this corresponds to the usual radial rescaling map. As in topology, this construction defines a naive $\mathbb{A}^1$-homotopy between the identity map ($t = 1$) and the map factoring through the inclusion of 0 ($t = 1$). In any case, affine space gives the first example of a space satisfying the hypotheses of the following definition.

Definition 5.3.1.1 A space $\mathscr{X} \in \mathrm{Spc}_k$ is $\mathbb{A}^1$-contractible if the structure morphism $\mathscr{X} \to \mathrm{Spec}\,k$ is an $\mathbb{A}^1$-weak equivalence.   □

Example 5.3.1.2 Assume k is a field and let α_1 and α_2 be coprime integers. The cuspidal curve $\Gamma_{\alpha_1,\alpha_2} = \{y^{\alpha_1} - z^{\alpha_2} = 0\}$ is $\mathbb{A}^1$-contractible. More precisely, we identify $\Gamma_{\alpha_1,\alpha_2}$ as a motivic space by restricting its functor of points to Sm_k. Then, the normalization map $\mathbb{A}^1_x \to \Gamma_{\alpha_1,\alpha_2}$ given by $x \mapsto (x^{\alpha_2}, x^{\alpha_1})$ is an $\mathbb{A}^1$-weak equivalence, even an isomorphism of presheaves on Sm_k (see [10, Example 2.1]).

□

Example 5.3.1.3 Over a perfect field k, there are no non-trivial forms of the affine line.   □

Suppose Z is an $\mathbb{A}^1$-contractible smooth scheme. Since we have forced maps that are "Nisnevich locally" weak equivalences to be weak equivalences, it follows that any map that is Nisnevich locally isomorphic to the product projection $U \times_{\mathrm{Spec}\,k} Z \to U$ is automatically a weak equivalence. More precisely, if $f : X \to Y$ is a morphism of smooth k-schemes, and there exists a Nisnevich covering map $u : U \to Y$ and an isomorphism of U-schemes $X \times_Y U \cong U \times_{\mathrm{Spec}\,k} Z$, then we will say that f is Nisnevich locally trivial with $\mathbb{A}^1$-contractible fibers.

Lemma 5.3.1.3 *If $f : X \to Y$ is any morphism of smooth k-schemes that is Nisnevich locally trivial with affine space fibers, then f is an $\mathbb{A}^1$-weak equivalence.*

□

For example, the projection map in a geometric vector bundle is automatically an $\mathbb{A}^1$-weak equivalence. More generally, if $\pi : E \to X$ is a torsor under a vector bundle on X, then π is Zariski locally trivial (this follows from the vanishing of coherent cohomology on affine schemes), and thus an $\mathbb{A}^1$-weak equivalence. Jouanolou originally observed [85] that given a quasi-projective variety X, one could find a torsor under a vector bundle over X whose total space was affine; such a scheme will be called an *affine vector bundle torsor*. Thomason generalized this observation to schemes admitting an ample family of line bundles; and in the next result we summarize the consequences for $\mathbb{A}^1$-homotopy theory.

Lemma 5.3.1.4 (Jouanolou–Thomason Homotopy Lemma) *If k is a regular ring, and X is a separated, finite type, smooth k-scheme, then there exists a smooth affine k-scheme $\tilde{X}$ and morphism $\pi : \tilde{X} \to X$ that is a torsor under a vector bundle; in particular, π is Zariski locally trivial with affine space fibers and is thus an $\mathbb{A}^1$-weak equivalence.*   □

Proof If k is regular, then X is a separated, regular, Noetherian scheme. In that case, X admits an ample family of line bundles (combine [136, Expose III Corollaire 2.2.7.1]. The result then follows from [148, Proposition 4.4], a result attributed to Thomason.   □

Remark 5.3.1.5 If X is a smooth scheme, then a choice of smooth affine scheme $\tilde{X}$ and an $\mathbb{A}^1$-weak equivalence $\pi : \tilde{X} \to X$ will be called a Jouanolou device over X. Unfortunately, the construction of Jouanolou devices is not functorial. □

Example 5.3.1.6 For any base ring k, following Jouanolou, there is a very simple example of a Jouanolou device over $\mathbb{P}^n$. Let $\widetilde{\mathbb{P}^n}$ be the complement of the incidence hyperplane in $\mathbb{P}^n \times \mathbb{P}^n$ (viewing the second $\mathbb{P}^n$ as the dual of the first). It is easy to see that the composite of the inclusion and the projection onto the first factor defines a morphism $\widetilde{\mathbb{P}^n} \to \mathbb{P}^n$ which is a Jouanolou device; we will refer to this as the standard Jouanolou device over $\mathbb{P}^n$. For $n = 1$ it is straighforward to check that $\widetilde{\mathbb{P}^1}$ is isomorphic to the closed subscheme of $\mathbb{A}^3_k$ defined by the equation $xy = z(1+z)$. □

5.3.2 $\mathbb{A}^1$-weak Equivalences vs. Weak Equivalences

For this section, we will consider rings k that come equipped with a homomorphism $\iota : k \to \mathbb{C}$. In that case, we may compare $\mathbb{A}^1$-weak equivalences and classical weak equivalences via what are often called "realization functors". Given a smooth k-scheme X, we may consider the set X^{an}_ι (via ι) and we view this as a complex manifold with its usual structure. Morel and Voevodsky [111, §3.3] show that the assignment $X \mapsto X^{an}_\iota$ may then be extended to a functor between homotopy categories

$$\mathfrak{R}_\iota : \mathscr{H}(k) \longrightarrow \mathcal{H}$$

which we refer to as a (topological) realization functor; see [49] for more discussion of topological realization functors. In particular, applying $\mathfrak{R}_\iota$ to an $\mathbb{A}^1$-weak equivalence of smooth schemes yields a weak equivalence of the associated topological spaces of complex points.

Remark 5.3.2.1 The choice ι is important: Serre showed that it is possible to find smooth algebraic varieties over a number field together with two embeddings of k into $\mathbb{C}$ such that the resulting complex manifolds are homotopy inequivalent. In fact, F. Charles provided examples of two smooth algebraic varieties over a number field k together with two embeddings of k into $\mathbb{C}$ such that the real cohomology algebras of the resulting complex manifolds are not isomorphic [32]. Said differently, the real homotopy type of a smooth k-scheme may depend on the choice of embedding of k into $\mathbb{C}$. □

Question 5.3.2.2 Assume $\iota_1, \iota_2 : k \to \mathbb{C}$ are distinct ring homomorphisms. Is it possible to find a (smooth) k-scheme X such that $\mathfrak{R}_{\iota_1}(X)$ is contractible while $\mathfrak{R}_{\iota_2}(X)$ is not? □

Remark 5.3.2.3 Recall that a connected topological space X is $\mathbb{Z}$-acyclic if $H_i(X, \mathbb{Z}) = 0$ for all $i > 0$. Of course, contractible topological spaces are $\mathbb{Z}$-acyclic. One can show that the property of being $\mathbb{Z}$-acyclic is independent of the choice of embedding for smooth varieties using étale cohomology as follows. If X is a smooth k-scheme, then the integral singular cohomology groups $H_i(X^{an}, \mathbb{Z})$ are finitely generated abelian groups. By the Artin-Grothendieck comparison theorem, the cohomology of $X(\mathbb{C})$ with $\mathbb{Z}/n$-coefficients is isomorphic to the étale cohomology of X with $\mathbb{Z}/n$-coefficients, and étale cohomology of X is independent of the choice of embedding of k into $\mathbb{C}$. By appeal to the universal coefficient theorem, one deduces that the vanishing of $H_i(X^{an}, \mathbb{Z})_{tors}$ for one embedding implies the vanishing for any other. Similarly, the rank of the free part is determined by the Betti numbers, which are also determined by étale cohomology and are therefore also independent of the choice of embedding.

To our knowledge, all the examples where homotopy types change with the embedding involve a nontrivial fundamental group. If X is a topologically contractible smooth k-scheme, then its étale fundamental group is independent of the choice of embedding. Furthermore, the étale fundamental group of X is the profinite completion of the topological fundamental group of $X(\mathbb{C})$. Thus, if X is topologically contractible, then any of the manifolds $X(\mathbb{C})$ has a fundamental group with trivial profinite completion. If one could prove that $X(\mathbb{C})$ has trivial fundamental group for any choice of embedding, the above problem would have a positive solution as a consequence of the usual Whitehead theorem. Let us also note that, working with étale homotopy types, one can deduce restrictions on the profinite completions of the other homotopy groups of $X(\mathbb{C})$.   □

Using the realization functor mentioned above and the definition of topologically contractible varieties from the introduction, the following result is immediate.

Lemma 5.3.2.4 *If k is a commutative ring, $\iota : k \to \mathbb{C}$ is a homomorphism, and X is $\mathbb{A}^1$-contractible smooth k-scheme, then $\mathfrak{R}_\iota(X)$ is topologically contractible.*   □

Remark 5.3.2.5 While complex realization is only available for fields admitting an embedding in $\mathbb{C}$, there are other realization functors that may be defined more generally. For example, one may define an étale realization functor on the unstable $\mathbb{A}^1$-homotopy category over any field [78]. Over fields having characteristic p, the p-part of the étale homotopy type is not $\mathbb{A}^1$-invariant in general. For example, the affine line over a separably closed field having positive characteristic has a large nontrivial étale fundamental group. Thus, étale realization of an unstable $\mathbb{A}^1$-homotopy type involves completion away from the residue characteristics of whatever base ring we work over. Correspondingly, the analog of Lemma 5.3.2.4 says that the étale realization of an $\mathbb{A}^1$-contractible smooth scheme is only trivial after a suitable completion. On the other hand, the true "topological" analog of contractibility, i.e., contractibility in the étale sense *including* triviality of the p-part is extremely restrictive: in fact, there are no nontrivial étale contractible varieties [71].   □

5.3.3   Cancellation Questions and $\mathbb{A}^1$-weak Equivalences

We now connect the discussion to the cancellation questions mentioned in the introduction: from the standpoint of $\mathbb{A}^1$-homotopy theory, this can be viewed as a source of many interesting $\mathbb{A}^1$-weak equivalences. Before discussing the biregular cancellation problem, we recall some results about the original (birational) Zariski cancellation problem. In the special case where X is a projective space, this question can be rephrased as follows: if X is a stably k-rational variety, then is it k-rational? The work of Beauville–Colliot-Thélène–Sansuc–Swinnerton-Dyer from the early 80s answered Zariski's original question in the negative [24], i.e., even over algebraically closed fields, there exist stably rational, non-rational varieties of dimension ≥ 3 (examples over algebraically closed fields cannot exist in dimension ≤ 2 by the classification of non-singular surfaces). The results following this development introduced a hierarchy of birational cancellation questions for discussion of which we refer the interested reader to [34, §1]. Correspondingly, we introduce a hierarchy of "biregular" cancellation questions mimicking the birational story.

Definition 5.3.3.1 Suppose X and Y are irreducible smooth k-schemes of the same dimension. Say that X and Y are

1. *stably isomorphic* if there exist an integer $n \geq 0$ such that $X \times \mathbb{A}^n \cong Y \times \mathbb{A}^n$;
2. *common direct factors* if there exists a smooth variety Z such that $X \times Z \cong Y \times Z$; and
3. *common retracts* if there exists a smooth variety Z and closed immersions $X \to Z$ and $Y \to Z$ admitting retractions. $\square$

Remark 5.3.3.2 If X and Y are stably isomorphic, one may ask for the smallest value of m for which $X \times \mathbb{A}^m \cong Y \times \mathbb{A}^m$; so it makes sense to refine stable isomorphisms and inquire about m-stable isomorphisms. $\square$

From the standpoint of $\mathbb{A}^1$-homotopy theory, this definition is important because of the following result.

Lemma 5.3.3.3 *Stably isomorphic smooth varieties are $\mathbb{A}^1$-weakly equivalent.* $\square$

Perhaps the original cancellation question, which was explicitly stated by Coleman and Enochs [33], asked whether 1-stably isomorphic affine varieties are isomorphic. More generally, Abhyanker–Heinzer–Eakin [1] asked whether stably isomorphic affine varieties are isomorphic. In [1], this question is introduced as follows: a ring A is called invariant if given a ring B and an isomorphism $A[x_1, \ldots, x_n] \cong B[y_1, \ldots, y_n]$ it follows that $A \cong B$. In fact, Abhyankar–Heinzer–Eakin proved that one-dimensional integral domains over a field are invariant [1, Theorem 3.3], i.e., cancellation holds for irreducible affine curves. Then, [1, Question 7.10] asks whether two-dimensional integral domains over a field are invariant, with particular attention drawn to the case of the affine plane. It is, of course, natural to consider the invariance question in higher dimensions as well.

The invariance question becomes more subtle as the dimension of varieties under consideration increases. In the early 1970s, Hochster gave the first counter-example to a cancellation problem over the real numbers [70], and similar examples were observed by M.P. Murthy (unpublished). In the mid 1970s, Iitaka and Fujita gave geometric conditions (non-negativity of the so-called logarithmic Kodaira dimension, an invariant taking values among $-\infty, 0, 1, 2, \ldots$) under which affine varieties that are common direct factors are isomorphic [77, Theorem 1]. From the point of $\mathbb{A}^1$-homotopy theory, what is more interesting is counter-examples to cancellation questions involving smooth varieties.

5.3.4 Danielewski Surfaces and Generalizations

In the late 1980s, Danielewski gave [36] a rather definitive counter-example to the invariance question for smooth affine surfaces: he wrote down smooth affine surfaces depending on a positive integer n, such that any two varieties in the class were stably isomorphic, and showed that the relevant varieties could be non-isomorphic for different values of n; we now discuss these varieties in detail.

Definition 5.3.4.1 Fix a polynomial $P(z)$ in one variable and an integer $n \geq 1$. The Danielewski surface $D_{n,P}$ is the closed subscheme of $\mathbb{A}^3$ defined by the equation $x^n y = P(z)$. □

Example 5.3.4.2 When $n = 1$ and $P(z) = z(1 + z)$, the variety $D_{n,P}$ is isomorphic to the standard Jouanolou device over $\mathbb{P}^1$ from Example 5.3.1.6. Assuming 2 is invertible in our base ring, it is also isomorphic to the standard hyperbolic quadric $xy + z^2 = 1$. □

If $P(z)$ is a separable polynomial, then it is straightforward to see that $D_{n,P}$ is smooth over k. In that case, projection in the x-direction determines a morphism $D_{n,P} \to \mathbb{A}^1_k$. Assume for simplicity k is an algebraically closed field, so $P(z)$ factors as a product of distinct linear factors; write $z_1, \ldots, z_d$ for the $d := deg P(z)$ distinct roots of $P(z)$. In that case, the fibers of the projection morphism are isomorphic to $\mathbb{A}^1_k$ over non-zero points of $\mathbb{A}^1_k$ while the fiber over 0 consists of d copies of $\mathbb{A}^1_k$ defined by $x = 0, z = z_i$. The complement of all but $d - 1$ of these copies of the affine line is an open subscheme of $D_{n,P}$ that is isomorphic to $\mathbb{A}^2$, and the restriction of the projection morphism is a product projection $\mathbb{A}^1_k \times \mathbb{A}^1_k \to \mathbb{A}^1_k$. Thus, the projection morphism $D_{n,P} \to \mathbb{A}^1_k$ factors through a morphism $D_{n,P} \to \mathbb{A}^1_P$, where $\mathbb{A}^1_P$ is a non-separated version of the affine line with a d-fold origin. Danielewski and Fieseler observed that this product projection makes $D_{n,P} \to \mathbb{A}^1_P$ into a torsor under a line bundle (see [40, Proposition 2.6] for various generalizations of this construction). We summarize the Danielewski construction in the following result, which follows from the discussion above and the fact that torsors under line bundles over affine schemes may be trivialized (i.e., are isomorphic to line bundles).

Proposition 5.3.4.3 (Danielewski, Fieseler, Dubouloz) *Assume k is algebraically closed and P is a separable polynomial. If n and n' are distinct positive integers, and P is a separable polynomial over k, then set:*

$$D_{n,P} \xleftarrow{p_1} D_{n,P} \times_{\mathbb{A}^1_P} D_{n',P} \xrightarrow{p_2} D_{n',P}.$$

The morphisms p_i make the fiber product into the projection map for a geometric line bundle. In particular, $D_{n,P}$ and $D_{n',P}$ are common retracts of $D_{n,P} \times_{\mathbb{A}^1_P} D_{n',P}$ (retraction given by the zero section). If $P = z(1 + z)$, then $D_{n,P}$ and $D_{n',P}$ are furthermore stably isomorphic. $\square$

The above observations, coupled with a homotopy colimit argument allow one to describe the $\mathbb{A}^1$-homotopy type of $D_{n,P}$ rather explicitly: it is $\mathbb{A}^1$-weakly equivalent to a wedge sum of $d - 1$-copies of $\mathbb{P}^1_k$. The proposition even gives very explicit $\mathbb{A}^1$-weak equivalences for different values of n and fixed P. The isomorphism class of the varieties $D_{n,z(1+z)}$ may be distinguished over the complex numbers by computing their first homology at infinity: explicitly D_n may be realized as the complement of a divisor in a Hirzebruch surface.

Proposition 5.3.4.4 (Fieseler) *If $k = \mathbb{C}$, then for any integer $n \geq 2$, $H_1^\infty(D_{n,z(1+z)}) \cong \mathbb{Z}/2n\mathbb{Z}$. In particular, if n and n' are distinct integers, $D_{n,z(1+z)}$ and $D_{n',z(1+z)}$ are not isomorphic.* $\square$

Danielewski's original construction has been expanded in many directions. We refer the reader to [40] and [41] for more details. One even knows that there are pairs of topologically contractible smooth affine varieties giving counter-examples to cancellation [46]. Cancellation may fail for open subsets of affine space: this was observed for affine spaces of sufficiently large dimension in [84] and for $\mathbb{A}^3$ in [43]. Jelonek even observed [84, Proposition 3.18] that there exist smooth affine varieties that are 2-stably isomorphic but not 1-stably isomorphic. Subsequently, lower-dimensional examples of this phenomena (though which fail to be smooth) were constructed in [7] and 2-dimensional smooth counterexamples were constructed in [44]. Furthermore, cancellation may fail rather generically: for every smooth affine variety of dimension $d \geq 7$, there exists a smooth affine variety X' birationally equivalent to X such that the variety $X' \times \mathbb{A}^2$ is not invariant. We refer the reader to [125] for a survey of the state of affairs up to about 2014, though the references above should make it clear that many exciting developments have occurred since that time.

Problem 5.3.4.5 Develop tools to distinguish isomorphism classes of smooth schemes having a given unstable $\mathbb{A}^1$-homotopy type. $\square$

Remark 5.3.4.6 In Sect. 5.4.4, we will develop some tools to aid in the study of this problem for smooth schemes that are not proper. $\square$

5.3.5   Building Quasi-Affine $\mathbb{A}^1$-contractible Varieties

Winkelmann's example 10 is realized as a quotient of affine space by a free action of a unipotent group (we review this below in Example 5.3.5.1. In this section, we discuss some examples that significantly expand on this idea and therefore show that $\mathbb{A}^1$-contractible smooth schemes are abundant in nature.

Unipotent Quotients

If the additive group scheme $\mathbb{G}_a$ acts scheme-theoretically freely on a smooth scheme X and a quotient exists as a scheme, then the quotient map is automatically Zariski locally trivial because $H^1(-, \mathbb{G}_a)$ vanishes on affine schemes ($\mathbb{G}_a$ is an example of a linear algebraic group that is *special* in the sense of Serre). In that case, the quotient map is automatically an $\mathbb{A}^1$-weak equivalence by appeal to Lemma 5.3.1.3.

Example 5.3.5.1 Take $k = \mathbb{Z}$, and suppress it from notation. Let Q_4 be the smooth affine quadric in $\mathbb{A}^5$ defined by the equation $x_1 x_3 - x_2 x_4 = x_5 (1 + x_5)$. Let E_2 be the closed subscheme defined by the equation $x_1 = x_3 = x_5 + 1 = 0$; E_2 is isomorphic to $\mathbb{A}^2$. The complement $X_4 := Q_4 \smallsetminus E_2$ is quasi-affine and not affine; it is Winkelmann's quasi-affine quotient from the introduction (i.e., Example 10) over $\operatorname{Spec} \mathbb{Z}$. The variety X_4 is an $\mathbb{A}^1$-contractible smooth scheme by appeal to Lemma 5.3.1.3.   □

Generalizing these observations, in [10], the first author and B. Doran showed that many (pairwise non-isomorphic) strictly quasi-affine $\mathbb{A}^1$-contractible varieties could be constructed in this way. In fact, the following result, which is a first step in the direction of Theorem 1.

Theorem 5.3.5.2 ([10, Theorem 1.3]) *Assume k is a field. For every integer $m \geq 6$ and every integer $n \geq 0$, there exists a connected n-dimensional k-scheme S and a smooth morphism $\pi : X \to S$ of relative dimension m, whose fibers over k-points are $\mathbb{A}^1$-contractible, quasi-affine, not affine, and pairwise non-isomorphic.*   □

By Proposition 5.2.4.6, it follows that if X is any $\mathbb{A}^1$-contractible smooth k-scheme, then $Pic(X) = Pic(k)$. In particular, if $Pic(k)$ is trivial (e.g., if k is a field or $\mathbb{Z}$), then every line bundle on X is trivial. In that case, every torsor under a line bundle on X is automatically a $\mathbb{G}_a$-torsor. It is known that total spaces of $\mathbb{G}_a$-torsors over schemes may have non-isomorphic total spaces (e.g., consider the Danielewski varieties above). The following question extends a question posed initially by Kraft [90, §3 Remark 2] in the case $X = X_4$ (since $\mathbb{G}_a$-torsors on affine schemes are always trivial, the question is only interesting for quasi-affine $\mathbb{A}^1$-contractible varieties).

Question 5.3.5.3 Suppose X is an $\mathbb{A}^1$-contractible smooth k-scheme. Is it possible to have two $\mathbb{G}_a$-torsors on X with non-isomorphic total spaces? □

Winkelmann's example also has interesting consequences for the shape of the generalized Serre question 6. Indeed, one may use it to see that it *is* possible to have nontrivial vector bundles on $\mathbb{A}^1$-contractible smooth schemes that are not affine. This phenomenon is analyzed in great detail in [11] as a measure of the failure of $\mathbb{A}^1$-invariance of the functor assigning to a scheme X the set of isomorphism classes of rank r vector bundles on X (cf. Example 5.2.4.7). It also shows that the affineness assertion in Question 6 is absolutely essential.

Example 5.3.5.4 The variety X_4 above carries a nontrivial rank 2 vector bundle. The variety Q_4 carries a nontrivial rank 2 vector bundle. The easiest way to see this is to realize Q_4 as $Sp_4/(Sp_2 \times Sp_2)$, i.e., as the quaternionic projective line $\mathbb{HP}^1$ in the sense of [114]. In that case, the map $Sp_4/Sp_2 \to Q_4$ is a nontrivial Sp_2-torsor, and the relevant vector bundle is the associated vector bundle to the standard 2-dimensional representation of this Sp_2-torsor. In fact, the quotient Sp_4/Sp_2 and the variety Q_4 both have the $\mathbb{A}^1$-homotopy type of a motivic sphere, and the relevant morphism is the motivic Hopf map sometimes called ν. Since the open immersion $j : X_4 \hookrightarrow Q_4$ has closed complement of codimension 2, the restriction functor j^* on the category of vector bundles is fully-faithful. Thus, this rank 2 bundle restricts to a nontrivial rank 2 vector bundle on X_4. The total space of this rank 2 vector bundle is another $\mathbb{A}^1$-contractible smooth scheme which is necessary non-isomorphic to affine space as it is itself quasi-affine! □

We have seen above that there are quasi-affine $\mathbb{A}^1$-contractible smooth schemes of dimension $d \geq 4$. On the other hand, one may see using classification that the only $\mathbb{A}^1$-contractible smooth scheme of dimension 1 (say over a perfect field) is $\mathbb{A}^1$. It follows from a general result of Fujita [54, §2 Theorem 1] that any topologically contractible smooth complex surface is necessarily affine. Thus, the following question remains open.

Question 5.3.5.5 If k is a field, does there exist an $\mathbb{A}^1$-contractible (resp. topologically contractible) smooth k-scheme of dimension 3 that is quasi-affine but not affine? □

Other Quasi-Affine $\mathbb{A}^1$-contractible Varieties

In [10], it was asked whether every quasi-affine $\mathbb{A}^1$-contractible variety could be realized as a quotient of an affine space by a free action of a unipotent group, generalizing the situation in topology suggested by Theorem 2. The answer to this question was seen to be no in [16]. For any integer $n \geq 0$, write Q_{2n} for the smooth affine k-scheme defined by the equation $\sum_i x_i y_i = z(1 + z)$ in $\mathbb{A}^{2n+1}$ with coordinates $(x_1, \ldots, x_n, y_1, \ldots, y_n, z)$. Then, define E_n to be the closed subscheme of Q_{2n} defined by $x_1 = \cdots = x_n = 1 + z = 0$. As in the case $n = 2$, E_n is isomorphic to affine space of dimension n. We define a variety

$$X_{2n} := Q_{2n} \setminus E_n.$$

For $n = 0$, $E_0 = \operatorname{Spec} k$, and X_0 is $\operatorname{Spec} k$ as well. For $n = 1$, one can check that X_2 is isomorphic to $\mathbb{A}_k^2$. For $n = 2$, X_4 is Winkelmann's example studied above. The following example shows that not every $\mathbb{A}^1$-contractible smooth scheme may be realized as a quotient of affine space by a free action of the additive group, in contrast to the situation in topology summarized in Theorem 2.

Theorem 5.3.5.6 ([16, Theorems 3.1.1 and Corollary 3.2.2]) *Suppose $n \geq 0$.*

1. *The variety X_{2n} is $\mathbb{A}^1$-contractible.*
2. *If $n \geq 3$, then X_{2n} is not a quotient of affine space by a free action of a unipotent group.*   □

Proof We will sketch the proof of the first statement, which follows by induction starting from the fact that $X_2 \cong \mathbb{A}^2$ and is thus $\mathbb{A}^1$-contractible. To set up the induction, we introduce some terminology. Let U_n be the open subscheme of Q_{2n} defined by $x_n \neq 0$. Note that U_n is isomorphic to $\mathbb{A}^{2n-1} \times \mathbb{G}_m$ with coordinates $x_1, \ldots, x_{n-1}, y_1, \ldots, y_n, z$ on the $\mathbb{A}^{2n-1}$-factor and coordinate x_n on the $\mathbb{G}_m$-factor. The closed complement Z_n of U_n, i.e., the closed subscheme of Q_{2n} defined by $x_n = 0$ is isomorphic to $Q_{2n-2} \times \mathbb{A}^1$ with coordinate y_n on the $\mathbb{A}^1$-factor. The point $x_1 = \cdots = x_{n-1} = y_1 = \cdots = y_n = z = 0$ defines a point 0 on Z_n. The normal bundle to Z_n is a trivial line bundle, with an explicit trivialization defined by the equation $x_n = 0$. Note that, by construction E_n is a closed subscheme of Z_n, and therefore $U_{2n} \subset X_{2n}$ as well. Likewise, the subscheme $Z_n \setminus E_n$ is isomorphic to $X_{2n-2} \times \mathbb{A}^1$.

The closed subscheme of U_n defined by setting $x_1 = \cdots = x_{n-1} = y_1 = \cdots = y_n = z = 0$ is isomorphic to $\mathbb{G}_m$ with coordinate x_n. The inclusion map $\mathbb{G}_m \to U_n$ is a monomorphism of presheaves, and splits the projection $U_n \cong \mathbb{A}^{2n-1} \times \mathbb{G}_m \to \mathbb{G}_m$. In particular, the map $U_n \to \mathbb{G}_m$ is an $\mathbb{A}^1$-weak equivalence. Likewise, the closed subscheme of Q_{2n} defined by $x_1 = \cdots = x_{n-1} = y_1 = \cdots = y_n = z = 0$ is isomorphic to $\mathbb{A}^1$ with coordinate x_n and we have a pullback diagram of the form

$$
\begin{array}{ccc}
0 & \longrightarrow & \mathbb{A}^1 \\
\downarrow & & \downarrow \\
Z_n & \longrightarrow & Q_{2n};
\end{array}
$$

where the top map is given by $x_n = 0$ and thus maps on normal bundles to the horizontal closed immersions have compatible trivializations. Since E_n is disjoint from this copy of $\mathbb{A}^1$, it follows that we have a sequence of inclusions of the form:

$$\mathbb{G}_m \longrightarrow \mathbb{A}^1 \longrightarrow X_{2n}.$$

We would like to understand the homotopy cofiber of the composite map, which coincides with the homotopy cofiber of the map $U_n \to X_n$ since the projection map $U_n \to \mathbb{G}_m$ is an $\mathbb{A}^1$-weak equivalence.

Since $\mathbb{A}^1$ is $\mathbb{A}^1$-contractible, the homotopy cofiber of the map $\mathbb{A}^1 \to X_{2n}$ is X_{2n} pointed by 0. The cofiber of $\mathbb{G}_m \to \mathbb{A}^1$ is $\mathbb{P}^1$ by the purity isomorphism (see Theorem 5.2.4.14). Likewise, the cofiber of the map $\mathbb{G}_m \to X_{2n}$ coincides with $Th(\nu_{(Z_n \smallsetminus E_n)/X_{2n}}) = \mathbb{P}^1_+ \wedge (X_{2n-2} \times \mathbb{A}^1)$, and there is thus a cofiber sequence of the form

$$\mathbb{P}^1 \longrightarrow \mathbb{P}^1_+ \wedge (Z_n \smallsetminus E_n) \longrightarrow X_{2n}.$$

By construction, and functoriality of the purity isomorphism (again Theorem 5.2.4.14), the left hand map is the $\mathbb{P}^1$-suspension of the map $S^0_k \to (Z_n \smallsetminus E_n)_+$ given by inclusion of 0 in the first factor. It is therefore split by the map $(Z_n \smallsetminus E_n)_+ \to S^0_k$ that corresponds to adding a disjoint base-point to the structure map. Thus, one concludes that there is an induced $\mathbb{A}^1$-weak equivalence $\mathbb{P}^1 \wedge (Z_n \smallsetminus E_n) \to X_{2n}$. Since $(Z_n \smallsetminus E_n) \cong X_{2n-2} \times \mathbb{A}^1$, it is $\mathbb{A}^1$-contractible, and the suspension $\mathbb{P}^1 \wedge (Z_n \smallsetminus E_n)$ is also $\mathbb{A}^1$-contractible. $\qquad\square$

Question 5.3.5.7 Is the total space of a Jouanolou device over X_{2n} isomorphic to an affine space? $\qquad\square$

Remark 5.3.5.8 An approach to analyzing the above question was suggested around 2016 in unpublished work of A. Ananyevskiy and A. Luzgarev [4], at least over fields having characteristic unequal to 2, and suggested independently by Danielewski [37] after reading an earlier version of this survey. The idea is to realize Q_{2n} as the homogeneous space SO_{2n+1}/SO_{2n} for the split special orthogonal group. In that case, one may try to build an explicit morphism from an affine space to X_{2n} as follows. Take the unipotent radical of a suitable parabolic subgroup of SO_{2n+1}, and a suitable subgroup of the unipotent radical of the opposite subgroup. In that case, one can use multiplication in SO_{2n+1} to build the relevant morphism. Nevertheless, in this case, it is not known if the morphism one obtains is Zariski (or Nisnevich) locally trivial. $\qquad\square$

5.4 Further Computations in $\mathbb{A}^1$-homotopy Theory

In the preceding section, we revisited some constructions for affine (and quasi-affine) varieties from the standpoint of $\mathbb{A}^1$-homotopy theory. Of these constructions and results, only Theorem 5.3.5.6 really required tools of $\mathbb{A}^1$-homotopy theory. To connect with some of the other questions mentioned in the introduction, in particular the generalized van de Ven Question 8, in this section we will discuss connectivity from the standpoint of $\mathbb{A}^1$-homotopy theory, and close with some of the basic computations of the analogs of classical (unstable) homotopy groups of spheres. Recall that k is a base Noetherian commutative unital ring of finite Krull dimension, which we will often assume is a field.

5.4.1 $\mathbb{A}^1$-homotopy Sheaves

Suppose $(\mathcal{X}, x)$ is a pointed space. Earlier, we defined bi-graded motivic spheres $S^{i,j}$. These bi-graded motivic spheres allow us to define corresponding homotopy groups.

Basic Definitions

Definition 5.4.1.1 Given a space $\mathcal{X}$, the sheaf of $\mathbb{A}^1$-connected components, denoted $\pi_0^{\mathbb{A}^1}(\mathcal{X})$, is the Nisnevich sheaf on Sm_k associated with the presheaf $U \mapsto [U, \mathcal{X}]_{\mathbb{A}^1}$. □

Definition 5.4.1.2 If $\mathcal{X}$ is a motivic space, then $\mathcal{X}$ is $\mathbb{A}^1$-connected if the canonical morphism $\pi_0^{\mathbb{A}^1}(\mathcal{X}) \to \mathrm{Spec}\, k$ is an isomorphism (and $\mathbb{A}^1$-disconnected otherwise). □

Definition 5.4.1.3 Given a pointed space $(\mathcal{X}, x)$, the i-th $\mathbb{A}^1$-homotopy sheaf, denoted $\pi_i^{\mathbb{A}^1}(\mathcal{X}, x)$, is the Nisnevich sheaf on Sm_k associated with the presheaf $U \mapsto [S^i \wedge U_+, (\mathcal{X}, x)]_{\mathbb{A}^1}$. □

As in classical topology, one can formally show that $\pi_1^{\mathbb{A}^1}(\mathcal{X}, x)$ is a Nisnevich sheaf of groups, and $\pi_i^{\mathbb{A}^1}(X, x)$ is a Nisnevich sheaf of abelian groups for $i > 1$. In fact, results of Morel show that, just like in topology, these sheaves of groups are "discrete" in an appropriate sense; see [109] for an introduction to these ideas and [110] for details. The following result, called the $\mathbb{A}^1$-Whitehead theorem for its formal similarity to the ordinary Whitehead theorem for CW complexes, is a formal consequence of the definitions.

Proposition 5.4.1.4 ([111, §3 Proposition 2.14]) A morphism $f : \mathcal{X} \to \mathcal{Y}$ of $\mathbb{A}^1$-connected spaces is an $\mathbb{A}^1$-weak equivalence if and only if for any choice of base-point x for X, setting $y = f(x)$ the induced morphism

$$\pi_i^{\mathbb{A}^1}(\mathcal{X}, x) \longrightarrow \pi_i^{\mathbb{A}^1}(\mathcal{Y}, y)$$

is an isomorphism. □

The following result is called the unstable 0-connectivity theorem.

Theorem 5.4.1.5 ([111, §2 Corollary 3.22]) If $\mathcal{X}$ is a space, then the canonical map $\mathcal{X} \to \pi_0^{\mathbb{A}^1}(\mathcal{X})$ is an epimorphism after Nisnevich sheafification. □

Remark 5.4.1.6 One consequence of Theorem 5.4.1.5 is that existence of a k-rational point is an $\mathbb{A}^1$-homotopy invariant. □

$\mathbb{A}^1$-rigid Varieties Embed into $\mathscr{H}(k)$

One rather fundamental difference between the $\mathbb{A}^1$-homotopy category and the classical homotopy category is that while classical homotopy types are essentially discrete, $\mathbb{A}^1$-homotopy types may vary in families. We begin by recalling the following definition, which begins to analyze the interaction with morphisms from the affine line and $\mathbb{A}^1$-connected components.

Definition 5.4.1.7 A smooth scheme of finite type $X \in \mathrm{Sm}_k$ is $\mathbb{A}^1$-rigid if the map

$$\mathrm{Sm}_k(Y \times \mathbb{A}^1, X) \longrightarrow \mathrm{Sm}_k(Y, X) \tag{5.1}$$

induced by the 0-section $\mathrm{Spec}\, k \to \mathbb{A}^1$ is a bijection for every $Y \in \mathrm{Sm}_k$. $\quad\square$

Remark 5.4.1.8 Let $\pi : Y \times \mathbb{A}^1 \to \mathbb{A}^1$ denote the projection map. Then the composite map

$$\mathrm{Sm}_k(Y, X) \xrightarrow{\pi^*} \mathrm{Sm}_k(Y \times \mathbb{A}^1, X) \longrightarrow \mathrm{Sm}_k(Y, X) \tag{5.2}$$

is the identity. Thus X is $\mathbb{A}^1$-rigid if and only if π^* is surjective or equivalently (5.1) is injective for all $Y \in \mathrm{Sm}_k$. $\quad\square$

Lemma 5.4.1.9 *A smooth scheme of finite type $X \in \mathrm{Sm}_k$ is $\mathbb{A}^1$-local fibrant if and only if it is $\mathbb{A}^1$-rigid.* $\quad\square$

Proof Since every object of Sm_k is local projective fibrant, $X \in \mathrm{Sm}_k$ is $\mathbb{A}^1$-local fibrant if and only if (5.1) is a weak equivalence of simplicial sets. We note that every discrete simplicial set is cofibrant and fibrant. Thus (5.1) is a weak equivalence if and only if it is a homotopy equivalence or equivalently a bijection. $\quad\square$

Corollary 5.4.1.10 *The full subcategory of $\mathbb{A}^1$-rigid schemes in Sm_k embeds fully faithfully into $\mathscr{H}(k)$. Moreover, if X is $\mathbb{A}^1$-rigid, the canonical morphism $X \to \pi_0^{\mathbb{A}^1}(X)$ is an isomorphism of Nisnevich sheaves.* $\quad\square$

Example 5.4.1.11 We note that $\mathbb{G}_m$ is $\mathbb{A}^1$-rigid. To show (5.1) is injective for all $Y \in \mathrm{Sm}_k$ we may assume $Y = \mathrm{Spec}\, R$, where R is a finitely generated k-algebra. In fact, if R is a reduced commutative ring, then the pullback map $R^\times \to R[x]^\times$ is bijective. More generally, any open subscheme of $\mathbb{G}_m$ is $\mathbb{A}^1$-rigid. Similarly, one may show that if X is a smooth curve of genus $g \geq 1$ and $U \subset X$ is an open subscheme, then U is also $\mathbb{A}^1$-rigid. $\quad\square$

Example 5.4.1.12 The scheme $\mathbb{P}^1$ is not $\mathbb{A}^1$-rigid because π^* is not surjective when $Y = \mathrm{Spec}\, k$ (there are of course many non-constant embeddings $\mathbb{A}^1 \hookrightarrow \mathbb{P}^1$). We will explore this example more in Sect. 5.4.2. $\quad\square$

Example 5.4.1.13 Recall that a semi-abelian variety is a smooth connected algebraic group G obtained as an extension

$$1 \longrightarrow T \longrightarrow G \longrightarrow A \longrightarrow 1$$

of an abelian variety A, i.e., a smooth connected proper algebraic group, by a torus T. For example, A can be the Jacobian of an algebraic curve of positive genus and T can be the multiplicative group scheme. For any map $\phi \colon \mathbb{A}^1 \to G$ the composite $\mathbb{A}^1 \to G \to A$ is constant. To show this we may assume k is algebraically closed. Indeed every map $\rho \colon \mathbb{P}^1 \to A$ is constant: by Lüroth's theorem, we are reduced to consider the normalization, i.e., we may assume that ρ is birational onto its image. In that case, the differential $d\rho \colon T_{\mathbb{P}^1} \to T_A$ is injective being non-zero at the generic point. Now the tangent sheaf T_A is trivial while $T_{\mathbb{P}^1} \cong \mathscr{O}_{\mathbb{P}^1}(2)$. However, there is no injective map $\mathscr{O}_{\mathbb{P}^1}(2) \to \mathscr{O}_{\mathbb{P}^1}^{\oplus n}$, where n is the dimension of A.

Returning to ϕ we conclude there exists $g \in G(k)$ such that ϕ factors through the translate $gT \subset G$, and we may view ϕ as a map $\mathbb{A}^1 \to (\mathbb{A}^1 \smallsetminus \{0\})^n$ for some $n > 0$. It follows that f is constant because every map $\mathbb{A}^1 \to \mathbb{A}^1 \smallsetminus \{0\}$ is constant. In fact, the affine line provides a useful geometric characterization of semi-abelian varieties by Brion [29, Proposition 5.4.5]: If G is a smooth connected algebraic group over a perfect field, then G is a semi-abelian variety if and only if every map $\mathbb{A}^1 \to G$ is constant.

Finally, since abelian varieties exist in non-constant families, we see $\mathbb{A}^1$-homotopy types exist in non-constant families. □

Remark 5.4.1.14 Over the complex numbers, an algebraic variety X is called *Brody* hyperbolic if every holomorphic map $\mathbb{C} \to X$ is constant and *algebraically hyperbolic* (also called *Mori hyperbolic*) if every algebraic morphism $\mathbb{A}^1 \to X$ is constant. It is not hard to show that Mori hyperbolic varieties are $\mathbb{A}^1$-rigid in the sense above. □

5.4.2 $\mathbb{A}^1$-*connectedness and Geometry*

Having explored varieties that were discrete from the standpoint of $\mathbb{A}^1$-homotopy theory, we now discuss aspects of connectedness in $\mathbb{A}^1$-homotopy theory. Recall that a motivic space $\mathscr{X}$ is $\mathbb{A}^1$-connected if the canonical morphism $\pi_0^{\mathbb{A}^1}(\mathscr{X}) \to \operatorname{Spec} k$ is an isomorphism (and $\mathbb{A}^1$-disconnected otherwise).

Remark 5.4.2.1 Since the $\mathbb{A}^1$-homotopy category is constructed by a localization procedure, the sheaf $\pi_0^{\mathbb{A}^1}(\mathscr{X})$ is rather abstractly defined and hard to "compute" in practice. To give one indication of how $\mathbb{A}^1$-connectedness interacts with arithmetic, suppose X is a k-scheme with k a field. Since stalks in the Nisnevich topology on Sm_k are henselizations of points on smooth schemes, Theorem 5.4.1.5 implies that if X is an $\mathbb{A}^1$-connected smooth scheme, then $X(S)$ is non-empty for S every henselization of a smooth scheme at a point. In particular, $\mathbb{A}^1$-connected smooth schemes always have k-rational points.   □

One would like to have a more "geometric" interpretation of $\mathbb{A}^1$-connectedness. Of course, any $\mathbb{A}^1$-contractible space is $\mathbb{A}^1$-connected, by the very definition. For this, we recall how connectedness is studied in topology: a topological space is *path* connected if any two points can be connected by a map from the unit interval. Replacing the unit interval by the affine line, we could define a notion of $\mathbb{A}^1$-path connectedness. For flexibility, we will use a slightly more general definition.

Definition 5.4.2.2 If X is a smooth k-scheme, say that X is $\mathbb{A}^1$-*chain connected* if for every separable, finitely generated extension K/k, $X(K)$ is non-empty, and for any pair $x, y \in X(K)$, there exist an integer N and a sequence $x = x_0, x_1, \ldots, x_N = y \in X(K)$ together with morphisms $f_1, \ldots, f_N : \mathbb{A}^1_K \to X$ with the property that $f_i(0) = x_{i-1}$ and $f_i(1) = x_i$; loosely speaking: any two points can be connected by the images of a chain of maps from the affine line.   □

Remark 5.4.2.3 Note: K is not necessarily a finite extension, so this definition is nontrivial even when $k = \mathbb{C}$. Indeed, in that case, we ask, e.g., that $\mathbb{C}(t)$-points, $\mathbb{C}(t_1, t_2)$-points, etc. can all be connected by the images of chains of affine lines.

□

From the definitions given, it is not clear that either $\mathbb{A}^1$-connectedness implies $\mathbb{A}^1$-chain connectedness or vice versa. In one direction, the problem is that $\mathbb{A}^1$-chain connectedness only imposes conditions over fields: while fields are examples of stalks in the Nisnevich topology, they do not exhaust all examples of stalks.

Proposition 5.4.2.4 ([107, Lemma 3.3.6] and [108, Lemma 6.1.3]) *If X is an $\mathbb{A}^1$-chain connected smooth variety, then X is $\mathbb{A}^1$-connected.*   □

Proof (Idea of Proof) The proof uses the fact that we are working with the Nisnevich topology in a fairly crucial way. To check triviality of all stalks, it suffices to show that $\pi_0^{\mathbb{A}^1}(X)(S)$ is trivial for S a henselian local scheme. Chain connectedness implies that the sections over the generic point of S are trivial and also that the sections over the closed point are trivial. We can then try to use a sandwiching argument to establish that sections over S are also trivial: in practice, this uses the homotopy purity Theorem 5.2.4.14!   □

Conversely, it is not clear that $\mathbb{A}^1$-connectedness implies $\mathbb{A}^1$-chain connectedness. However, one may prove the following result.

Theorem 5.4.2.5 ([15, Theorem 6.2.1]) *If X is a smooth proper k-variety, and K/k is any separable finitely generated extension, then $\pi_0^{\mathbb{A}^1}(X)(K) = X(K)/ \sim$. In particular, if X is $\mathbb{A}^1$-chain connected, then X is $\mathbb{A}^1$-connected.*   □

Remark 5.4.2.6 Another proof of this result has been given in [22].   □

$\mathbb{A}^1$-connectedness and Rationality Properties

The preceding theorem suggests a link between $\mathbb{A}^1$-connectedness and rationality properties. Indeed, Manin defined the notion of R-equivalence for rational points on an algebraic variety: if L/k is a finite extension and X is a smooth k-scheme, we say two L-points x_0 and x_1 are directly R-equivalent if there exists a rational map $\mathbb{P}^1_L \to X_L$ defined at the points x_0 and x_1. We say two L-points are R-equivalent if they are equivalent for the equivalence relation generated by direct R-equivalence, and we write $X(L)/R$ for the set of R-equivalence classes of L-rational points. One says that a variety X/k is *universally R-trivial* if $X(L)/R = *$ for every finitely generated separable extension L/k. With that definition, a smooth proper variety X that is $\mathbb{A}^1$-chain connected is automatically *universally R-trivial*, and Theorem 5.4.2.5 may be phrased as saying that $\mathbb{A}^1$-connected smooth proper varieties are universally R-trivial. In fact, one may make a slightly stronger version of this statement.

Proposition 5.4.2.7 *If k is a field and U is an $\mathbb{A}^1$-connected smooth k-variety that admits a smooth proper compactification X, then X is $\mathbb{A}^1$-connected and U is universally R-trivial.*   □

Remark 5.4.2.8 Of course, the hypothesis of admitting a smooth proper compactification is superfluous if either k has charateristic 0 or U has small dimension.   □

Proof The easiest proof of this fact uses a tool that we have not yet introduced called the zeroth $\mathbb{A}^1$-homology sheaf, which plays a role very similar to singular homology in classical topology, but is rather far from motivic (co)homology mentioned earlier. Since $U \subset X$ is open, the map

$$\mathrm{H}_0^{\mathbb{A}^1}(U) \longrightarrow \mathrm{H}_0^{\mathbb{A}^1}(X).$$

is an epimorphism by Asok [9, Proposition 3.8]. On the other hand, since U is $\mathbb{A}^1$-connected, there is an isomorphism $\mathrm{H}_0^{\mathbb{A}^1}(U) \cong \mathbb{Z}$. Since X is proper, one concludes $\mathrm{H}_0^{\mathbb{A}^1}(X) \cong \mathbb{Z}$ as well. Then, the fact that X is $\mathbb{A}^1$-connected follows from [9, Theorem 4.15]. To conclude, simply observe that if x_0 and x_1 are L-points in U, then x_0 and x_1 are connected by a chain of affine lines over L in X. Restricting this chain of affine lines to U gives the required witness to R-equivalence.   □

Recall that if k is a field, a k-form of the affine line is a smooth k-scheme of dimension 1 that is fppf-locally isomorphic to $\mathbb{A}^1_k$. However, if k is imperfect, then there exist non-trivial forms of the affine line (e.g., if $a \in k$ is not a p-th power, then the hypersurface in $\mathbb{A}^2_k$ defined by the equation $y^p = x + ax^p$ gives a non-trivial form).

Proposition 5.4.2.9 *If k is a field, and X is a form of the affine line over k, then X is $\mathbb{A}^1$-contractible if and only if X is isomorphic to $\mathbb{A}^1_k$.*   □

Proof One direction is immediate. Assume k is a field and fix an algebraic closure $\bar{k}$ of k. If $X_{\bar{k}}$ is not $\mathbb{A}^1$-connected, then X cannot be $\mathbb{A}^1$-connected either. Using this observation and Proposition 5.4.2.7 one easily reduces to the assertion that if X is $\mathbb{A}^1$-contractible, it must be isomorphic to a form of the affine line. If k is a perfect field, then there are no non-trivial forms of the affine line (see, e.g., [124, Lemma 1.1]), and the result follows.

Assume then that k is imperfect. R. Achet proves [2, Theorem 2.4] that if X is a non-trivial form of $\mathbb{A}^1_k$, then either $X(k)$ is empty or $Pic(X)$ is non-trivial. Note that both conclusions are eliminated by the assertion that X is $\mathbb{A}^1$-contractible. □

In light of the van de Ven question mentioned in the introduction, we observe that Proposition 5.4.2.7 has the following consequence on the rationality of $\mathbb{A}^1$-contractible varieties.

Corollary 5.4.2.10 *Assume k is a field having characteristic 0. If X is an $\mathbb{A}^1$-contractible smooth k-scheme then X is universally R-trivial.* □

Remark 5.4.2.11 If k is a field and X is a smooth proper variety, then any $\mathbb{A}^1$-connected smooth proper variety is separably rationally connected. However, $\mathbb{A}^1$-connectedness has cohomological implications, e.g., the (prime to the characteristic part of the) Brauer group of an $\mathbb{A}^1$-connected smooth proper k-scheme is automatically trivial (see [9, §4] for a detailed discussion of this point). For example, it is known that there exist k-unirational varieties that are not $\mathbb{A}^1$-connected [15, §2.3] [9, Example 4.18]. Thus, $\mathbb{A}^1$-connectedness of a smooth scheme has nontrivial implications for the rationality properties of the scheme. □

We know that $\mathbb{A}^1$-connectedness implies universal R-triviality for smooth schemes admitting a smooth proper compactification. However, the example of $\mathbb{G}_m$ shows that R-equivalent k-points need not be connected by (chains of) rational curves. Nevertheless, the following question remains open:

Question 5.4.2.12 Is it true for an arbitrary smooth k-scheme X that $\mathbb{A}^1$-connectedness is equivalent to $\mathbb{A}^1$-chain connectedness? □

Remark 5.4.2.13 The relationship between R-equivalence and $\mathbb{A}^1$-weak equivalence of points has been studied on certain linear group schemes in [21]. □

5.4.3 $\mathbb{A}^1$-homotopy Sheaves Spheres and Brouwer Degree

Earlier, we saw that $\mathbb{A}^n \smallsetminus 0$ was a motivic sphere: it was isomorphic in $\mathcal{H}(k)$ to $S^{n-1,n}$. It is not hard to show that $\mathbb{A}^n \smallsetminus 0$ is $\mathbb{A}^1$-connected for $n \geq 2$, so it is natural to inquire about its connectivity and to compute its first non-vanishing $\mathbb{A}^1$-homotopy sheaf. We quickly summarize some results of F. Morel, though we do not have enough space to motivate the proofs.

Theorem 5.4.3.1 (F. Morel) *If k is a field, then $\mathbb{A}^n \smallsetminus 0$ is at least $(n - 2)$-$\mathbb{A}^1$-connected, i.e., if $n \geq 2$, it is $\mathbb{A}^1$-connected and $\pi_i^{\mathbb{A}^1}(\mathbb{A}^n \smallsetminus 0, x)$ vanishes for any choice of k-point $x \in \mathbb{A}^n \smallsetminus 0(k)$ and any integer $1 \leq i \leq n - 2$.* $\qquad\square$

Morel also computed the first non-vanishing $\mathbb{A}^1$-homotopy sheaf in terms of Milnor–Witt K-theory introduced earlier.

Theorem 5.4.3.2 (F. Morel) *If k is a field, then for any integer $n \geq 2$, and any finitely generated separable extension L/k*

$$\pi_{n-1}^{\mathbb{A}^1}(\mathbb{A}^n \smallsetminus 0)(L) = \mathbf{K}_n^{MW}(L).$$

From this result, Morel deduced the computation of the homotopy endomorphisms of $\mathbb{A}^n \smallsetminus 0$. If $k \hookrightarrow \mathbb{C}$, then $\mathbb{A}^n(\mathbb{C})$ is homotopy equivalent to S^{2n-1}. Therefore, realization defines a homomorphism $[\mathbb{A}^n \smallsetminus 0, \mathbb{A}^n \smallsetminus 0]_{\mathbb{A}^1} \to [S^{2n-1}, S^{2n-1}] \cong \mathbb{Z}$ where the latter identification is the usual Brouwer degree.

Theorem 5.4.3.3 *If k is a field, then for any integer $n \geq 2$, there is a canonical "motivic Brouwer degree" isomorphism*

$$[\mathbb{A}^n \smallsetminus 0, \mathbb{A}^n \smallsetminus 0]_{\mathbb{A}^1} \cong GW(k).$$

Moreover, if $k \hookrightarrow \mathbb{C}$, the induced homomorphism $GW(k) \to \mathbb{Z}$ coincides with the homomorphisms induced by sending a stable isomorphism class of symmetric bilinear forms to the rank of its underlying $\mathbb{C}$-vector space. $\qquad\square$

Later, we will see that computations of motivic Brouwer degree appear in proofs of $\mathbb{A}^1$-contractibility statements.

5.4.4 $\mathbb{A}^1$-homotopy at Infinity

In this section, we introduce some notions of $\mathbb{A}^1$-homotopy theory at infinity. Unfortunately, we are unable at the moment to make a good definition of "end space" in order to define a workable notion of $\mathbb{A}^1$-fundamental group at infinity.

One-point Compactifications

Fix a field k, and suppose X is a smooth k-scheme. By a smooth compactification of X we will mean a pair $(\bar{X}, \psi)$, where $\psi : X \to \bar{X}$ is an open dense immersion, and $\bar{X}$ is smooth. We will say that such a smooth compactification is *good* if the closed complement of ψ (viewed as a scheme with the reduced induced structure) is a simple normal crossings divisor.

Lemma 5.4.4.1 *If X is a smooth scheme, and $\bar{X}_0$ and $\bar{X}_1$ are smooth compactifications, with boundaries $\partial\bar{X}_0$ and $\partial\bar{X}_1$, then $\bar{X}_0/\partial\bar{X}_0$ and $\bar{X}_1/\partial\bar{X}_1$ are cdh-locally weak equivalent and thus $\Sigma\bar{X}_0/\partial\bar{X}_0$ and $\Sigma\bar{X}_1/\partial\bar{X}_1$ are weakly equivalent in the $\mathbb{A}^1$-homotopy category.* □

Proof By definition of the cdh topology, if we are given an abstract blow-up square of the form

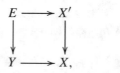

i.e., $Y \to X$ is a closed immersion, $X' \to X$ is proper and the induced map $X' \smallsetminus E \to X \smallsetminus Y$ is an isomorphism, then there is a cdh local weak equivalence $X'/E \to X/Y$. To establish the statement, simply take the closure $\bar{X}$ of the image of the diagonal map $X \to \bar{X}_0 \times \bar{X}_1$ and observe that $\bar{X} \to \bar{X}_0$ and $\bar{X} \to \bar{X}_1$ yield abstract blow-up squares. The second statement follows from the first because any morphism of presheaves that is a cdh local weak equivalence becomes a Nisnevich local weak equivalence after a single suspension (i.e., Theorem 5.2.4.16). □

The lemma above shows that one-point compactifications are well-behaved in the cdh-local version of the $\mathbb{A}^1$-homotopy category. Alternatively, the S^1-stable $\mathbb{A}^1$-homotopy type of a one-point compactification is well-defined.

Definition 5.4.4.2 If X is a smooth k-scheme, then for any compactification $\bar{X}$ of X, we set $\dot{X} = \bar{X}/\partial\bar{X}$; there is a natural map $X \to \dot{X}$. □

Lemma 5.4.4.3 *Suppose X is a smooth scheme. The following statements hold.*

1. *If X is proper, then $\dot{X} = X_+$.*
2. *For any integer $n \geq 0$, there is an $\mathbb{A}^1$-weak equivalence $(X \times \mathbb{A}^n) \cong \dot{X} \wedge \mathbb{P}^{1\wedge n}$.*

□

Proof The first statement is immediate from $X/\emptyset = X_+$. For the second statement, one can first observe that $\mathbb{P}^n$ is a compactification of $\mathbb{A}^n$ with boundary $\mathbb{P}^{n-1}$ and use the $\mathbb{A}^1$-weak equivalence $\mathbb{P}^n/\mathbb{P}^{n-1} \cong \mathbb{P}^{1\wedge n}$. Then, use the fact that if X is any compactification of X, then $\bar{X} \times \mathbb{P}^n$ is a compactification of $X \times \mathbb{A}^n$. □

Stable End Spaces

Our goal is to make some progress toward a definition of end-spaces in algebraic geometry. There are many possible approaches to such a definition, and we do not know whether they are all equivalent. One approach is to consider a "punctured formal neighborhood" of the boundary in a compactification; this approach is not suited to motivic homotopy theory using smooth schemes. The following definition is motivated by the definition of singular homology at infinity, in the case where the boundary is suitably "tame".

J. Wildeshaus introduced the notion of "boundary motive" [152] of a variety; it can be thought of as a motivic version of the singular chain complex at infinity. In fact, with notational modifications, Wildeshaus' definition gives a $\mathbb{P}^1$-stable homotopy type. The only novelty of the definition we give below is that it gives an S^1-stable homotopy type.

Definition 5.4.4.4 Assume X is a smooth k-scheme. The S^1-stable end space is defined to fit into the following exact triangle:

$$X \coprod \{\infty\} \longrightarrow \dot{X} \longrightarrow e(X)[1]$$

Remark 5.4.4.5 The main benefit of working S^1-stably instead of $\mathbb{P}^1$-stably is that one has some hope of uncovering unstable phenomena related to "$\mathbb{A}^1$-connected components at ∞". Indeed, the zeroth $\mathbb{A}^1$-homology sheaf of a smooth proper scheme (see the proof of Proposition 5.4.2.7) detects rational points, while the corresponding $\mathbb{P}^1$-stable object only sees zero cycles of degree 1 [9, 14]. Thus, the above definition should be refined enough to detect some algebro-geometric analog of the number of ends of a space.   □

Proposition 5.4.4.6 *Assume k is a field having characteristic 0.*

1. *The construction $e(-)$ is a functor on the category of smooth schemes and proper maps.*
2. *If X is S^1-stably $\mathbb{A}^1$-contractible, then $e(X) \cong \dot{X}/\infty[-1]$.*   □

Example 5.4.4.7 The end space of $\mathbb{R}$ is S^0. Analogously, $e(\mathbb{A}^1) = \mathbb{G}_m$. In particular, end spaces need not even be $\mathbb{A}^1$-connected. Similarly, we find $e(\mathbb{A}^n) \cong \mathbb{A}^n \setminus 0$.   □

Example 5.4.4.8 Suppose X is an $\mathbb{A}^1$-contractible smooth affine scheme of dimension n. If X is isomorphic to $\mathbb{A}^n$ then $e(X) \cong \mathbb{A}^n \setminus 0$.   □

End spaces as defined here are also compatible with realization. Indeed, the triangle defining $e(X)$ makes sense in the usual stable homotopy category, and we define $e_{\mathbb{C}}(X)$ to be the cofiber of the natural map from the suspension spectrum of $X(\mathbb{C}) \coprod \{\infty\}$ to the suspension spectrum of $\dot{X}(\mathbb{C})$. Because realization behaves well with respect to homotopy cofibers, the following result is immediate.

Proposition 5.4.4.9 *If $\iota : k \hookrightarrow \mathbb{C}$ is an embedding, then $\mathfrak{R}_\iota(e(X)) = e_{\mathbb{C}}(X)$.*   □

Question 5.4.4.10 Is there a "good" unstable definition of end spaces in $\mathbb{A}^1$-homotopy theory?   □

5.5   Cancellation Questions and $\mathbb{A}^1$-contractibility

5.5.1   The Biregular Cancellation Problem

After the work of Iitaka–Fujita and Danielewski, it became clear that cancellation questions could admit negative solutions for smooth affine varieties of negative logarithmic Kodaira dimension. Perhaps the main remaining question in this direction is as follows.

Question 5.5.1.1 (Biregular Cancellation) If X is a smooth scheme of dimension d such that

1. X is stably isomorphic to $\mathbb{A}^d$, or
2. X is a direct factor of $\mathbb{A}^N$ for some $N > d$, or
3. X is a retract of $\mathbb{A}^N$ for some $N > d$, or

is X necessarily isomorphic to affine space?   □

Remark 5.5.1.2 The question of whether varieties that are stably isomorphic to affine space are necessarily isomorphic to affine space is sometimes called *the Zariski cancellation question*, but as mentioned earlier, Zariski never explicitly stated this question. On the other hand Beilinson apparently asked whether any retract of affine space is isomorphic to affine space [156, §8].   □

The biregular cancellation question is known to have a positive answer for smooth affine schemes of dimension 1 by Abhyankar et al. [1], and also for smooth affine schemes of dimension 2: the result was established by Miyanish–Sugie and Fujita [53, 104] over characteristic 0 fields and extended to perfect fields of arbitrary characteristic in [121]. In contrast, we now know that the biregular cancellation problem admits a negative answer over algebraically closed fields having positive characteristic. Indeed, N. Gupta constructed a counter-example in dimension 3 [59] and extended this result in a number of directions [60, 61]. We refer the reader to [62] for more discussion of these results. However, the specific form of these counterexamples does not allow them to be lifted to characteristic 0.

Lemma 5.5.1.3 *If X is a smooth scheme of dimension d that is a retract of $\mathbb{A}^N$, then X is a smooth affine, $\mathbb{A}^1$-contractible scheme. In particular, any counter-example to the biregular cancellation problem is necessarily a smooth affine $\mathbb{A}^1$-contractible scheme.*   □

Proof Any retract of an $\mathbb{A}^1$-weak equivalence is an $\mathbb{A}^1$-weak equivalence.   □

Remark 5.5.1.4 Gupta's counterexamples to cancellation above provided the first examples of smooth *affine* $\mathbb{A}^1$ contractible schemes in positive characteristic.   □

Remark 5.5.1.5 Over $\mathbb{C}$, the biregular cancellation question remains open in dimensions ≥ 3.   □

Granted the Quillen–Suslin theorem on triviality of vector bundles on affine space, it is easy to see that any algebraic vector bundle on a variety that is a retract of affine space is automatically trivial. The representability theorem for vector bundles guarantees that the same statement holds for $\mathbb{A}^1$-contractible smooth affine varieties.

Theorem 5.5.1.6 *If X is a smooth affine $\mathbb{A}^1$-contractible variety, then all vector bundles on X are trivial.* $\square$

5.5.2 $\mathbb{A}^1$-contractibility vs Topological Contractibility

We now compare $\mathbb{A}^1$-contractibility and topological contractibility in more detail. In particular, we would like to know whether topological contractibility and $\mathbb{A}^1$-contractibility are actually different. The best we can do at the moment is to proceed dimension by dimension. The only topologically contractible smooth curve is $\mathbb{A}^1$. However, already in dimension 2 problems appear to arise.

Affine Lines on Topologically Contractible Surfaces

With the exception of $\mathbb{A}^2$, most topologically contractible surfaces appear to have very few affine lines, as we now explain. Based on the classification results (see [156] and the references therein for more details), it suffices to treat the case of surfaces of logarithmic Kodaira dimensions 1 and 2 (there are no contractible surfaces of logarithmic Kodaira dimension 0).

Remark 5.5.2.1 General conjectures in algebraic geometry and arithmetic of Green–Griffiths and Lang suggest that smooth proper varieties of general type should not have "many" rational curves (see, e.g., [39]). Analogously, one hopes that affine varieties of log general type should not have "many" morphisms from the affine line (see, e.g., [97]). The topologically contractible surfaces of logarithmic Kodaira dimension 2 contain no contractible curves by work of Zaidenberg [154, 155] and Miyanishi-Tsunoda [105]; what can one say about morphisms from the affine line to such a surface? For example, are such surfaces Mori hyperbolic (see Remark 5.4.1.14)? $\square$

Remark 5.5.2.2 The surfaces of logarithmic Kodaira dimension 1 are all obtained from some special surfaces (the so-called tom Dieck-Petrie surfaces) by repeated application of a procedure called an affine modification (an affine variant of a blow-up). How does $\mathbb{A}^1$-chain connectedness behave with respect to affine modifications (we understand well how $\mathbb{A}^1$-chain connectedness behaves with respect to blow-ups of projective schemes with smooth centers)? One could also try to use the rationality results of Gurjar-Shastri, i.e., that any smooth compactification of a topologically contractible surface is rational [64, 65]. $\square$

Based on these observations, it seems reasonable to expect that topologically contractible surfaces that are not isomorphic to $\mathbb{A}^2$ are disconnected from the standpoint of $\mathbb{A}^1$-homotopy theory, which leads to the following conjecture suggesting an answer to Question 12 from the introduction.

Conjecture 5.5.2.3 A smooth topologically contractible surface X is $\mathbb{A}^1$-contractible if and only if it is isomorphic to $\mathbb{A}^2$.   □

The generalized van de Ven Question 8 asks whether all topologically contractible varieties are rational. For $\mathbb{A}^1$-contractible varieties, by "soft" methods, one can establish "near rationality" as we observed above. The upshot of this discussion is that $\mathbb{A}^1$-contractibility is a significantly stronger restriction on a space than topological contractibility. In support of the above conjecture, the following classification result was observed in [47].

Proposition 5.5.2.4 *An $\mathbb{A}^1$-contractible and $\mathbb{A}^1$-chain connected smooth affine surface over an algebraically closed field of characteristic 0 is isomorphic to the affine plane $\mathbb{A}^2$.*   □

Example 5.5.2.5 ([47]) For coprime integers $k > l \geq 2$, the smooth tom Dieck-Petrie surface is defined as

$$\mathcal{V}_{k,l} := \{ \frac{(xz + 1)^k - (yz + 1)^l - z}{z} = 0\} \subset \mathbb{A}^3 \tag{5.3}$$

We note that $\mathcal{V}_{k,l}$ is topologically contractible and stably $\mathbb{A}^1$-contractible. However, $\mathcal{V}_{k,l}$ has logarithmic Kodaira dimension $\overline{\kappa}(\mathcal{V}_{k,l}) = 1$, and thus it cannot be $\mathbb{A}^1$-chain connected. This example shows that the affine modification construction does not preserve $\mathbb{A}^1$-chain connectedness. It is an open question whether or not $\mathcal{V}_{k,l}$ is $\mathbb{A}^1$-contractible.   □

Establishing $\mathbb{A}^1$-connectedness is a first step towards understanding $\mathbb{A}^1$-homotopy type. A first step toward answering Question 8 in higher dimensions thus seems to begin with an analysis of the following problem.

Problem 5.5.2.6 Which classes of topologically contractible varieties are known to be $\mathbb{A}^1$-chain connected?   □

Chow Groups and Vector Bundles on Topologically Contractible Surfaces

The only general result about vector bundles smooth topologically contractible varieties of arbitrary dimension pertains to the Picard group.

Theorem 5.5.2.7 ([63, Theorem 1]) *If X is a topologically contractible smooth complex variety, then $Pic(X) = 0$.*   □

Proof Gurjar states this result for affine varieties. To remove the affineness assumption, one may either inspect the proof and see that the assumption is never used, or one may reduce to the affine case by simply observing that $Pic(X)$ is $\mathbb{A}^1$-invariant by Proposition 5.2.4.6, and any topologically contractible smooth complex variety is $\mathbb{A}^1$-weakly equivalent to a topologically contractible smooth affine scheme by appeal to Lemma 5.3.1.4. □

On the other hand, Theorem 5.2.4.8 together with standard techniques of obstruction theory allows one to understand vector bundles on topologically contractible smooth affine varieties. Indeed, the ideas of [12, 13] show that classification of vector bundles on smooth affine schemes of low dimensions are reduced to analysis of Chow groups. If X is a topologically contractible smooth affine complex variety of dimension d, then one may relate the structure of $CH^d(X)$ to geometry. Indeed, if Y is any smooth complex affine variety of dimension d, then a theorem of Roitman [120] implies that $CH^d(Y)$ is uniquely divisible. Thus, if $CH^d(Y)$ is furthermore finitely generated, it must be trivial. The latter condition may be guaranteed by imposing conditions on the geometry of compactifications and is thus related to the generalized van de Ven Question 8. The Chow groups of topologically contractible surfaces, which were originally computed by Gurjar and Shastri [64, 65], may be computed in this way. Indeed, Gurjar and Shastri show that the generalized van de Ven question has a positive answer in dimension 2, so it follows immediately that if X is a topologically contractible smooth complex affine surface, then $CH^2(X)$ is trivial.

Theorem 5.5.2.8 *If X is a topologically contractible smooth complex surface, then every algebraic vector bundle on X is trivial.* □

Proof Suppose X is a smooth affine surface over an algebraically closed field. By Serre's splitting theorem, it suffices to prove that rank 1 and rank 2 bundles are trivial. However, $Pic(X) \cong CH^1(X)$. The results of [12, Theorem 1] imply that the canonical map

$$(c_1, c_2) : \mathscr{V}_2(X) \longrightarrow CH^1(X) \times CH^2(X)$$

is a bijection. If $k = \mathbb{C}$ and X is furthermore topologically contractible, then Theorem 5.5.2.7 implies that $CH^1(X) = Pic(X) = 0$. The argument that $CH^2(X) = 0$ is given before the statement. □

Remark 5.5.2.9 In fact, the results of Gurjar and Shastri give a much more refined result than $CH^2(X) = 0$ for a topologically contractible smooth complex affine variety. In [8], the Voevodsky motive of a topologically contractible smooth complex affine surface is seen to be isomorphic to that of a point. This implies that the Chow groups are *universally trivial*, i.e., for every finitely generated extension $L/\mathbb{C}$, $CH^2(X_L) = 0$ as well.

Similarly, the generalized Serre Question 6 in dimension 3 may be reduced to a question that is purely cohomological.

Theorem 5.5.2.10 *If X is a topologically contractible smooth complex threefold, then every algebraic vector bundle on X is trivial if and only if $CH^2(X)$ and $CH^3(X)$ are trivial.*   □

Proof As in the proof of Theorem 5.5.2.8, it suffices to treat the case of ranks 1, 2 and 3 by Serre's splitting theorem. It is known that the Picard group of any topologically contractible threefold is trivial, so the rank 1 case follows. The results of [12, Theorem 1] imply that if, furthermore, $CH^2(X)$ is trivial, then every rank 2 vector bundle is trivial. Finally, classical results of Mohan Kumar and Murthy imply that there is a unique rank 3 vector bundle with given (c_1, c_2, c_3) [106]. In particular, if $CH^3(X)$ is also trivial, it follows that every rank 3 bundle on such an X is trivial.   □

Remark 5.5.2.11 In [8], it is observed that one may produce threefolds with trivial Chow groups by means of the technique of affine modifications, so one may produce many examples of topologically contractible threefolds satisfying the above hypotheses. Similar observations are used in [74] to establish triviality of vector bundles on Koras–Russell threefolds (see Sect. 5.5.3 and the discussion after Theorem 5.5.3.7 for more details).   □

As sketched above, one may use geometry to analyze vector bundles on topologically contractible smooth complex affine threefolds. Indeed, suppose X is a topologically contractible smooth complex affine threefold admitting a smooth compactification $\bar{X}$ that is rationally connected. Of course, this is weaker than assuming the generalized van de Ven Question 8 has a positive solution in dimension 3. In that case $CH^3(\bar{X}) = \mathbb{Z}$ and thus $CH^3(X)$ is trivial. Tian and Zong [137, Theorem 1.3] implies that $CH^2(\bar{X})$ is generated by classes of rational curves. Using this, and the localization sequence, one can sometimes establish that $CH^2(X)$ is itself torsion. In that case, [91, Appendix Theorem] due to Srinivas implies that $CH^2(X)$ is actually trivial. For example, it follows from [91, Corollary 18] that if X is a topologically contractible smooth complex affine threefold that admits a finite morphism from $\mathbb{A}^3$, then all vector bundles on X are trivial.

Remark 5.5.2.12 For some comments on the situation in dimension ≥ 4, see Conjecture 5.5.3.11.   □

5.5.3   Cancellation Problems and the Russell Cubic

We now investigate $\mathbb{A}^1$-contractible smooth affine varieties over fields having characteristic 0. For concreteness, it's useful to focus on one particular case: let $\mathcal{KR}$ be the so-called Russell cubic, i.e., the smooth variety in $\mathbb{A}^4$ defined by the equation:

$$x + x^2 y + z^3 + t^2 = 0$$

There is a natural $\mathbb{G}_m$-action on $\mathcal{KR}$ given by

$$\mathbb{G}_m \times \mathcal{KR} \to \mathcal{KR}; \; (\lambda, x, y, z, t) \mapsto (\lambda^6 x, \lambda^{-6} y, \lambda^2 z, \lambda^3 t). \tag{5.1}$$

With respect to this action we have the $\mathbb{G}_m$-invariant variables $u := xy, v := yt^2 \in \mathcal{O}(\mathcal{KR})^{\mathbb{G}_m}$. In fact, the GIT-quotient for (5.1) is given by

$$\pi : \mathcal{KR} \to \mathbb{A}^2_{u,v}; \; (x, y, z, t) \mapsto (u, v). \tag{5.2}$$

The fiber $\pi^{-1}(\alpha, \beta) \subset \mathcal{KR}$ is described by the equations $\alpha = xy, \beta = yt^2$, and

$$x + x^2 y + z^3 + t^2 = 0. \tag{5.3}$$

Multiplying (5.3) by y yields the equation

$$\alpha + \alpha^2 + yz^3 + \beta = 0. \tag{5.4}$$

Using these equations one checks that there is exactly one closed orbit in each fiber. Thus the set of closed orbits is parameterized by the normal scheme $\mathbb{A}^2_{u,v}$ and we conclude the corresponding GIT-quotient is indeed (5.2).

Makar-Limanov succeeded in showing that $\mathcal{KR}$ is non-isomorphic to $\mathbb{A}^3$ [98] by calculating his eponymous invariant. We recall that the Makar-Limanov invariant of an affine algebraic variety X is the subring $\mathrm{ML}(X)$ of $\Gamma(X, \mathcal{O}_X)$ comprised of regular functions that are invariant under all $\mathbb{G}_a$-actions on X. Using the bijection between $\mathbb{G}_a$-actions on X and locally nilpotent derivations ∂ on the k-algebra $\Gamma(X, \mathcal{O}_X)$ one finds that

$$\mathrm{ML}(X) = \bigcap_\partial \ker(\partial).$$

Clearly we have $\mathrm{ML}(\mathbb{A}^3) = k$ and similarly for all affine spaces, while extensive calculations reveal that $\mathrm{ML}(\mathcal{KR}) = k[x]$. That is, $\mathcal{KR}$ admits in a sense fewer $\mathbb{G}_a$-actions than $\mathbb{A}^3$. Here we observe the inclusion $\mathrm{ML}(\mathcal{KR}) \subset k[x]$: the locally nilpotent derivations $x^2 \partial_z - 2z \partial_y$ and $x^2 \partial_t - 3t^2 \partial_y$ of $k[x, y, z, t]$ induce locally nilpotent derivations on the coordinate ring $k[\mathcal{KR}]$. One easy checks that their kernels intersect in $k[x]$. The interesting part of Makar-Limanov's calculation is to show that $\partial(x) = 0$ for every locally nilpotent derivation of $k[\mathcal{KR}]$. Alternatively one can use Kaliman's result in [86] saying that if the general fibers of a regular function $\mathbb{A}^3 \to \mathbb{A}^1$ are isomorphic to $\mathbb{A}^2$ then all its fibers are isomorphic to $\mathbb{A}^2$. All the closed fibers of the projection map $\mathcal{KR} \to \mathbb{A}^1_x$ are isomorphic to $\mathbb{A}^2$ except for over the origin, which yields a copy of the cylinder on the cuspidal curve $\{z^3 + t^2 = 0\}$.

Dubouloz [42] showed that the Makar-Limanov invariant cannot distinguish between the cylinder $\mathcal{KR} \times \mathbb{A}^1$ on the Russell cubic and the affine space $\mathbb{A}^4$.

Furthermore, M.P. Murthy showed that all vector bundles on $\mathcal{KR}$ are trivial [112] (it is also known that the Chow groups of $\mathcal{KR}$ are trivial). However, the $\mathbb{G}_m$-action on $\mathcal{KR}$ has an isolated fixed point. If $\mathcal{KR}$ is not stably isomorphic to $\mathbb{A}^3$, is there an $\mathbb{A}^1$-homotopic obstruction to stable isomorphism?

Question 5.5.3.1 Is the Russell cubic $\mathcal{KR}$ $\mathbb{A}^1$-contractible?   □

This question, which has recently been solved in the affirmative, has guided much of the research in the area. First, one might try to compute the $\mathbb{A}^1$-homotopy groups; for this even to be sensible, we should make sure that the first obstruction to $\mathbb{A}^1$-contractibility vanishes. For a generalization of the following observation we refer the reader to [47].

Proposition 5.5.3.2 (B. Antieau (unpublished)) *The Russell cubic $\mathcal{KR}$ is $\mathbb{A}^1$-chain connected.*   □

5.5.3.3 (Approach 1) Can one detect nontriviality of any of the higher $\mathbb{A}^1$-homotopy groups of $\mathcal{KR}$? One approach to this problem is to think "naively" of, e.g., the $\mathbb{A}^1$-fundamental group. Think of chains of maps from $\mathbb{A}^1$ that start and end at a fixed point up to "naive" homotopy equivalence (this naturally forms a monoid rather than a group). The resulting object maps to the actual $\mathbb{A}^1$-fundamental group, but what can one say about its image?   □

5.5.3.4 (Approach 2) Since to disprove $\mathbb{A}^1$-contractiblity, we only need one cohomology theory that is $\mathbb{A}^1$-representable that detects nontriviality, it is useful to look at invariants that are not as "universal" as $\mathbb{A}^1$-homotopy groups. For another approach, using group actions, let us mention that J. Bell showed that rational $\mathbb{G}_m$-equivariant K_0 of $\mathcal{KR}$ is actually nontrivial [25]. Unfortunately, his computations together with the Atiyah-Segal completion theorem in equivariant algebraic K-theory also show that the "Borel style" equivariant K_0 is isomorphic to the Borel style equivariant K_0 of a point [18]. Nevertheless, $\mathbb{A}^1$-homotopy theory gives a wealth of new cohomology theories with which to study the Russell cubic. For example, it would be interesting to know if one of the more "refined" Borel style equivariant theories is refined enough to detect failure of $\mathbb{A}^1$-contractibility.

If $\mu_n \subset \mathbb{G}_m$ is a sufficiently "large" subgroup then the μ_n-equivariant K_0 of $\mathcal{KR}$ is also nontrivial. Moreover, the fixed-point loci for the μ_n-actions are all affine spaces (indeed, if n is prime, then the only nontrivial subgroup is the trivial subgroup which has the total space as fixed point locus). Thus, for many purposes, one might simply look at μ_n-equivariant geometry.

In equivariant topology, a map is a "fine" equivariant weak equivalence if it induces a weak equivalence on fixed point loci for all subgroups. Transplanting this to $\mathbb{A}^1$-homotopy theory: if we knew that equivariant algebraic K-theory was representable in an appropriate equivariant $\mathbb{A}^1$ homotopy category, such a category has been constructed for finite groups by Voevodsky [38], and we knew enough about the weak equivalences, then Bell's result might formally imply that X is not $\mathbb{A}^1$-contractible.   □

The Russell Cubic and Equivariant K-theory

For representability of equivariant algebraic K-theory let us work relative to a regular Noetherian commutative unital ring k of finite Krull dimension. We assume that $G \to k$ is a finite constant group scheme (or more generally that it satisfies the resolution property: every coherent G-module on X in Sch_k^G is the equivariant quotient of a G-vector bundle). Under these assumptions, Nisnevich descent for equivariant algebraic K-theory of smooth schemes over k was established in [68]. The fact that equivariant algebraic K-theory satisfies equivariant Nisnevich descent for smooth schemes implies that it is representable in the equivariant motivic homotopy category.

Let X be a G-scheme over k. Write $\mathcal{P}^G(X)$ for the exact category of G-vector bundles. Then the equivariant algebraic K-groups are the homotopy groups $K_i^G(X) := \pi_i \mathcal{K}(\mathcal{P}^G(X))$ of the associated K-theory space, defined by Waldhausen's $S_\bullet$-construction [147]. We obtain a presheaf of simplicial sets $\mathcal{K}^G$ on Sch_k^G such that $\pi_i \mathcal{K}^G(X) = K_i^G(X)$ for all X by applying a rectification procedure to the pseudo-functor $X \mapsto \mathcal{P}^G(X)$. With the same hypothesis as above, there is a natural isomorphism

$$K_i^G(X) \cong [S^i \wedge X_+, \mathcal{K}^G]_{\mathcal{H}_\bullet^G(k)}$$

for any X in Sm_k^G and the pointed G-equivariant motivic homotopy category $\mathcal{H}_\bullet^G(k)$ of k. This is the desired representability result for equivariant algebraic K-theory mentioned above.

An explicit computation of the μ_p-equivariant Grothendieck groups of Koras–Russell threefolds was carried out in [74]. For concreteness we specialize to the case $k = \mathbb{C}$. When X is a complex variety with an action of an algebraic group G, we let $R(G) \simeq K_0^G(k)$ denote the representation ring of G. If $H \subseteq G$ is a closed subgroup, we note there is a restriction map $K_0^G(X) \to K_0^H(X)$. Let X be a smooth affine variety with $\mathbb{C}^\times$-action and let $n > 0$ be an integer. There is a natural ring isomorphism

$$\phi \colon K_0^{\mathbb{C}^\times}(X) \underset{R(\mathbb{C}^\times)}{\otimes} R(\mu_n) \xrightarrow{\cong} K_0^{\mu_n}(X). \tag{5.5}$$

An algebraic $\mathbb{C}^\times$-action on a smooth complex affine variety is called hyperbolic if it has a unique fixed point and the weights of the induced linear action on the tangent space at this fixed point are all non-zero and their product is negative. Recall from [89] that a *Koras–Russell threefold* X is a smooth hypersurface in $\mathbb{A}_{\mathbb{C}}^4$ which is

1. topologically contractible,
2. has a hyperbolic $\mathbb{C}^\times$-action, and
3. the quotient $X//\mathbb{C}^\times$ is isomorphic to the quotient of the linear $\mathbb{C}^\times$-action on the tangent space at the fixed point (in the sense of GIT).

It is shown in [89, Theorem 4.1] that the coordinate ring of a threefold X satisfying (1)–(3) has the form

$$\mathbb{C}[X] = \frac{\mathbb{C}[x, y, z, t]}{t^{\alpha_2} - G(x, y^{\alpha_1}, z^{\alpha_3})}. \tag{5.6}$$

Here $\alpha_1, \alpha_2, \alpha_3$ are pairwise coprime positive integers. We let r denote the x-degree of the polynomial $G(x, y^{\alpha_1}, 0)$ and set $\epsilon_X = (r - 1)(\alpha_2 - 1)(\alpha_3 - 1)$. A Koras–Russell threefold X is said to be *nontrivial* if $\epsilon_X \neq 0$.

Bell [25] showed that the $\mathbb{C}^\times$-equivariant Grothendieck group of X is of the form

$$
\begin{aligned}
K_0^{\mathbb{C}^\times}(X) &= R(\mathbb{C}^\times) \oplus \left(\frac{R(\mathbb{C}^\times)}{(f(t))}\right)^{\rho-1} \\
&= \mathbb{Z}[t, t^{-1}] \oplus \left(\frac{\mathbb{Z}[t, t^{-1}]}{(f(t))}\right)^{\rho-1} \\
&= \mathbb{Z}[t, t^{-1}] \oplus \mathbb{Z}^{(\alpha_2-1)(\alpha_3-1)},
\end{aligned} \tag{5.7}
$$

where

$$f(t) = \frac{(1 - t^{\alpha_2 \alpha_3})(1 - t)}{(1 - t^{\alpha_2})(1 - t^{\alpha_3})} \tag{5.8}$$

is a polynomial of degree $(\alpha_2 - 1)(\alpha_3 - 1)$ and $\rho \geq 2$ is the number of irreducible factors of $G(x, y^{\alpha_1}, 0) \in \mathbb{C}[x, y]$. In particular, $K_0^{\mathbb{C}^\times}(X)$ is nontrivial. A combination of (5.5)–(5.8) together with explicit calculations reveal that the μ_p-equivariant Grothendieck group of X is trivial for almost all primes p. This implies that the suggested approach to showing non-$\mathbb{A}^1$-contractibility of a Koras–Russell threefold via μ_p-equivariant Grothendieck groups cannot work.

Theorem 5.5.3.5 *Let p be a prime and let $n \geq 1$ be an integer. Let μ_{p^n} act on a Koras–Russell threefold X via the inclusion $\mu_{p^n} \subset \mathbb{C}^\times$. Then the following hold.*

1. *The structure map $X \to \operatorname{Spec}(\mathbb{C})$ induces an isomorphism $R(\mu_{p^n}) \oplus F_{p^n} \xrightarrow{\cong} K_0^{\mu_{p^n}}(X)$.*
2. *F_{p^n} is a finite abelian group which is nontrivial if and only if X is nontrivial and $p \mid \alpha_2 \alpha_3$.* $\qquad\square$

If the integers p and $\alpha_2 \alpha_3$ are coprime it follows that every μ_{p^n}-equivariant vector bundle on X is stably trivial, i.e., for any μ_{p^n}-equivariant vector bundle E on X, there exist μ_{p^n}-representations F_1 and F_2 such that $E \oplus F_1 \simeq F_2$.

Higher Chow Groups and Stable $\mathbb{A}^1$-contractibility

Another natural idea is to study the higher Chow groups of Koras–Russell three-folds. Showing triviality of the said groups goes a long way in concluding $\mathbb{A}^1$-contractibility of $\mathcal{KR}$.

Proposition 5.5.3.6 *Let X be a Koras–Russell threefold of the first kind with coordinate ring*

$$\mathbb{C}[X] = \frac{\mathbb{C}[x, y, z, t]}{(ax + x^m y + z^{\alpha_2} + t^{\alpha_3})},$$

where $m > 1$ is an integer, $a \in \mathbb{C}^*$, and $\alpha_2, \alpha_3 \geq 2$ are coprime. For Y any smooth complex affine variety, the pullback map $CH^*(Y) \to CH^*(X \times Y)$ is an isomorphism. □

A related calculation shows the same conclusion holds for Koras–Russell threefolds of the second kind. As for the proof of Proposition 5.5.3.6 a key input is the observation that the ring homomorphism

$$\mathbb{C} \to \mathbb{C}[u, v]/(u^a + v^b)$$

induces an isomorphism on higher Chow groups for coprime integers $a, b \geq 2$.

Combined with the isomorphism between higher Chow groups and motivic cohomology, as shown by Voevodsky [142, Corollary 2], we obtain the following.

Theorem 5.5.3.7 *Let X be a Koras–Russell threefold of the first or second kind, and let Y be any smooth complex affine variety. Then the pullback map $H^{*,*}(Y, \mathbb{Z}) \to H^{*,*}(X \times Y, \mathbb{Z})$ induced by the projection $X \times Y \to Y$ is an isomorphism of (bigraded) integral motivic cohomology rings.* □

A consequence of Theorem 5.5.3.7 is that every vector bundle on X is trivial. This was originally shown by Murthy [112, Corollary 3.8] by a completely different method.

Theorem 5.5.3.7 is a key input in the approach to $\mathbb{A}^1$-contractibility of Koras–Russell threefolds in [74]. To proceed it is convenient to employ some techniques from stable motivic homotopy theory. In particular, the slice filtration on the stable motivic homotopy category $\mathscr{SH}(\mathbb{C})$ will be put to good use [143]. Recall the objects of $\mathscr{SH}(\mathbb{C})$ are sequences of pointed motivic spaces related by structure maps with respect to $(\mathbb{P}^1, \infty)$. We recall that $\mathscr{SH}(\mathbb{C})$ is a triangulated category with shift functor $E \mapsto E[1]$ given by smashing with the topological circle. Denote by $\Sigma_{\mathbb{P}^1}^\infty(X, x) \in \mathscr{SH}(\mathbb{C})$ the $(\mathbb{P}^1, \infty)$-suspension spectrum of $X \in \mathrm{Sm}_{\mathbb{C}}$ and a rational point $x \in X(\mathbb{C})$. For fixed $F \in \mathscr{SH}(\mathbb{C})$, we say that $E \in \mathscr{SH}(\mathbb{C})$ is

1. *F-acyclic* if $E \wedge F \simeq *$;
2. *F-local* if $\mathrm{Hom}_{\mathscr{SH}(\mathbb{C})}(D, E) = 0$ for every F-acyclic spectrum D.

It is clear that the F-local spectra form a colocalizing subcategory of $\mathscr{S}\mathscr{H}(\mathbb{C})$. Note that if F is a ring spectrum, then any F-module E is F-local (every map $D \to E$ factors through $D \wedge F$ and hence it is trivial if D is F-acyclic).

Let $\mathbf{MZ} \in \mathscr{S}\mathscr{H}(\mathbb{C})$ denote the motivic ring spectrum that represents motivic cohomology, i.e., for every $X \in \mathrm{Sm}_{\mathbb{C}}$ and integers $n, i \in \mathbb{Z}$ there is an isomorphism

$$H^{n,i}(X, \mathbb{Z}) \simeq \mathrm{Hom}_{\mathscr{S}\mathscr{H}(\mathbb{C})}(\Sigma_{\mathbb{P}^1}^\infty X_+, \mathbf{MZ}(i)[n]). \tag{5.9}$$

Here, for $E \in \mathscr{S}\mathscr{H}(\mathbb{C})$, the Tate twist $E(1)$ is defined by $E(1) = E \wedge \Sigma_{\mathbb{P}^1}^\infty(\mathbb{G}_m, 1)[-1]$. The Betti realization of $\mathbf{MZ}$ identifies with the classical Eilenberg-Mac Lane spectrum $\mathbf{HZ}$ representing singular (co)homology of topological spaces [69, Appendix A] and [94, §5].

Lemma 5.5.3.8 *For every* $X \in \mathrm{Sm}_{\mathbb{C}}$ *and closed point* $x \in X$ *the suspension* $\Sigma_{\mathbb{P}^1}^\infty(X, x) \in \mathscr{S}\mathscr{H}(\mathbb{C})$ *is* $\mathbf{MZ}$-*local.*   □

Proof Resolution of singularities allows one to show that $\Sigma_{\mathbb{P}^1}^\infty(X, x)$ is in the smallest thick subcategory of $\mathscr{S}\mathscr{H}(\mathbb{C})$ containing $\Sigma_{\mathbb{P}^1}^\infty Y_+$ for any smooth projective variety Y. It suffices now to show that $\Sigma_{\mathbb{P}^1}^\infty Y_+$ is $\mathbf{MZ}$-local for such Y. Voevodsky's slice filtration for any $E \in \mathscr{S}\mathscr{H}(\mathbb{C})$ is a tower of spectra [123, 143]

$$\cdots \to f_{q+1}E \to f_q E \to f_{q-1}E \to \cdots \to E, \quad q \in \mathbb{Z}.$$

Here the qth slice $s_q E$ of E is defined by the distinguished triangle

$$f_{q+1}E \to f_q E \to s_q E \to f_{q+1}E[1].$$

Levine [93] has shown that the slice filtration of $\Sigma_{\mathbb{P}^1}^\infty Y_+$ for Y any smooth projective variety is complete in the sense that

$$\underset{q \to \infty}{\mathrm{holim}}\, f_q(\Sigma_{\mathbb{P}^1}^\infty Y_+) \simeq *.$$

Equivalently, if we define $c_q E$ by the distinguished triangle $f_q E \to E \to c_q E \to f_q E[1]$, then

$$\Sigma_{\mathbb{P}^1}^\infty Y_+ \simeq \underset{q \to \infty}{\mathrm{holim}}\, c_q(\Sigma_{\mathbb{P}^1}^\infty Y_+).$$

Since the subcategory of $\mathbf{MZ}$-local spectra is colocalizing, it now suffices to prove that $c_q(\Sigma_{\mathbb{P}^1}^\infty Y_+)$ is $\mathbf{MZ}$-local for every $q \in \mathbb{Z}$. By definition of the slice filtration, we have $c_q(\Sigma_{\mathbb{P}^1}^\infty Y_+) \simeq *$ for $q \leq 0$. Using the distinguished triangles

$$s_q E \to c_q E \to c_{q-1}E \to s_q E[1]$$

and induction on q, we are reduced to proving the slices $s_q(\Sigma_{\mathbb{P}^1}^\infty Y_+)$ are **MZ**-local. In fact, all slices in $\mathscr{SH}(\mathbb{C})$ are **MZ**-local: any slice $s_q E$ is a module over the zeroth slice $s_0(\mathbf{1})$ of the sphere spectrum, and hence it is $s_0(\mathbf{1}) \simeq$ **MZ**-local [66, §6 (iv), (v)]. $\square$

Theorem 5.5.3.9 *Let X be a Koras–Russell threefold of the first or second kind. Then there exists an integer $n \geq 0$ such that the suspension $\Sigma_{\mathbb{P}^1}^n(X, 0)$ is $\mathbb{A}^1$-contractible.* $\square$

Proof We first reformulate Theorem 5.5.3.7 as an equivalence in $\mathscr{SH}(\mathbb{C})$, using its structure of a closed symmetric monoidal category, see Sect. 5.2.5. The structure map $X \to \mathrm{Spec}(\mathbb{C})$ induces a morphism in $\mathscr{SH}(\mathbb{C})$

$$\mathbf{MZ} \simeq \mathcal{H}om(\Sigma_{\mathbb{P}^1}^\infty \mathrm{Spec}(\mathbb{C})_+, \mathbf{MZ}) \to \mathcal{H}om(\Sigma_{\mathbb{P}^1}^\infty X_+, \mathbf{MZ}). \tag{5.10}$$

In view of (5.9), Theorem 5.5.3.7 asserts that for every smooth complex affine variety Y and $n, i \in \mathbb{Z}$, there is an induced isomorphism

$$\mathrm{Hom}_{\mathscr{SH}(\mathbb{C})}(\Sigma_{\mathbb{P}^1}^\infty Y_+(i)[n], \mathbf{MZ}) \to \mathrm{Hom}_{\mathscr{SH}(\mathbb{C})}(\Sigma_{\mathbb{P}^1}^\infty Y_+(i)[n], \mathcal{H}om(\Sigma_{\mathbb{P}^1}^\infty X_+, \mathbf{MZ})).$$

The objects $\Sigma_{\mathbb{P}^1}^\infty Y_+(i)[n]$ form a family of generators of $\mathscr{SH}(\mathbb{C})$, because every smooth variety admits an open covering by smooth affine varieties. Thus (5.10) and its retraction $\mathcal{H}om(\Sigma_{\mathbb{P}^1}^\infty X_+, \mathbf{MZ}) \to \mathbf{MZ}$ induced by the base point $0 \in X$ are isomorphisms. From the distinguished triangle

$$\mathbf{MZ}[-1] \to \mathcal{H}om(\Sigma_{\mathbb{P}^1}^\infty(X, 0), \mathbf{MZ}) \to \mathcal{H}om(\Sigma_{\mathbb{P}^1}^\infty X_+, \mathbf{MZ}) \to \mathbf{MZ},$$

we deduce that $\mathcal{H}om(\Sigma_{\mathbb{P}^1}^\infty(X, 0), \mathbf{MZ}) \simeq *$. By [119, Theorems 1.4 and 2.2] or [122, Theorem 52], $\Sigma_{\mathbb{P}^1}^\infty(X, 0)$ is strongly dualizable in $\mathscr{SH}(\mathbb{C})$, so that

$$\mathcal{H}om(\Sigma_{\mathbb{P}^1}^\infty(X, 0), \mathbf{MZ}) \simeq \mathcal{H}om(\Sigma_{\mathbb{P}^1}^\infty(X, 0), \mathbf{1}) \wedge \mathbf{MZ}.$$

Thus $\mathcal{H}om(\Sigma_{\mathbb{P}^1}^\infty(X, 0), \mathbf{1})$ is **MZ**-acyclic, and for any $E \in \mathscr{SH}(\mathbb{C})$ we obtain

$$\mathrm{Hom}_{\mathscr{SH}(\mathbb{C})}(E, \Sigma_{\mathbb{P}^1}^\infty(X, 0) \wedge \mathbf{MZ}) \simeq \mathrm{Hom}_{\mathscr{SH}(\mathbb{C})}(E \wedge \mathcal{H}om(\Sigma_{\mathbb{P}^1}^\infty(X, 0), \mathbf{1}), \mathbf{MZ}) \simeq *,$$

since $E \wedge \mathcal{H}om(\Sigma_{\mathbb{P}^1}^\infty(X, 0), \mathbf{1})$ is **MZ**-acyclic and **MZ** is **MZ**-local (being an **MZ**-module). By the Yoneda lemma, this implies $\Sigma_{\mathbb{P}^1}^\infty(X, 0)$ is **MZ**-acyclic, i.e.,

$$\Sigma_{\mathbb{P}^1}^\infty(X, 0) \wedge \mathbf{MZ} \simeq *,$$

On the other hand, by Lemma 5.5.3.8, $\Sigma_{\mathbb{P}^1}^\infty(X, 0)$ is **MZ**-local. It follows that every endomorphism of $\Sigma_{\mathbb{P}^1}^\infty(X, 0)$ is trivial, and hence $\Sigma_{\mathbb{P}^1}^\infty(X, 0) \simeq *$. Owing to Lemma 5.2.5.1 this completes the proof. $\square$

Remark 5.5.3.10 In general, $\mathbb{A}^1$-weak equivalences do not desuspend. To illustrate this, for simplicity, take $k = \mathbb{C}$, and let $\{p_1, \ldots, p_n\}$ and $\{q_1, \ldots, q_n\}$ be two collections of complex points in $\mathbb{A}^1$, but the argument works much more generally. If $n \geq 1$, then $\mathbb{A}^1 \smallsetminus \{p_1, \ldots, p_n\}$ is an $\mathbb{A}^1$-rigid variety in the sense of Definition 5.4.1.7. In particular, the $\mathbb{A}^1$-weak equivalences $\mathbb{A}^1 \smallsetminus \{p_1, \ldots, p_n\} \cong \mathbb{A}^1 \smallsetminus \{q_1, \ldots, q_n\}$ are simply isomorphisms of varieties by appeal to Corollary 5.4.1.10. However, any isomorphism of varieties of this form is induced by an automorphism of the affine line. The automorphism group of the affine line acts 2-transitively, but not n-transitively for any $n \geq 3$. In fact, as soon as $n \geq 3$, there is a moduli space of configurations of dimension $n - 2$.

The variety $\Sigma_{\mathbb{P}^1}\mathbb{A}^1 \smallsetminus \{p_1, \ldots, p_n\}$ is $\mathbb{A}^1$-weakly equivalent to $\mathbb{A}^2 \smallsetminus \{x_1, \ldots, x_n\}$ for any collection of n $\mathbb{C}$-points of $\mathbb{A}^2$. Indeed, it is not hard to show that the algebraic automorphism group of $\mathbb{A}^2$ acts n-fold transitively on $\mathbb{A}^2(\mathbb{C})$ for any $n \geq 1$, in contrast to the situation for the affine line. Thus, if we choose coordinates x, y on $\mathbb{A}^2$, we may move the points $x_1, \ldots, x_n$ to lie on the x-axis and then cover $\mathbb{A}^2 \smallsetminus \{x_1, \ldots, x_n\}$ by $\mathbb{A}^1 \smallsetminus \{p_1, \ldots, p_n\} \times \mathbb{A}^1$ and $\mathbb{A}^1 \times \mathbb{G}_m$ with intersection $\mathbb{A}^1 \smallsetminus \{p_1, \ldots, p_n\} \times \mathbb{G}_m$. The required weak equivalence then follows by the same homotopy colimit argument used to prove that $\mathbb{A}^2 \smallsetminus 0 \cong \mathbb{P}^1 \wedge \mathbb{G}_m$. By increasing the number of points, we see that there are arbitrary dimensional moduli of smooth varieties that become $\mathbb{A}^1$-weakly equivalent after a single $\mathbb{P}^1$-suspension. □

The discussion above also has implications for topologically contractible smooth complex varieties. The motivic conservativity conjecture (see, e.g., [75, Proposition 3.4] or [19, Conjecture 2.1']) implies the rational Voevodsky motive of a topologically contractible variety is that of a point. In dimension 2, the *integral* Voevodsky motive of topologically contractible surface is trivial by the results of [8]; it is also observed there that triviality holds for a number of higher dimensional examples. It thus is not inconsistent with known examples to suggest that the integral Voevodsky motive of a topologically contractible smooth complex variety is always that of a point. In conjunction with the proof of Theorem 5.5.3.9, the following seems reasonable.

Conjecture 5.5.3.11 If X is a topologically contractible smooth complex affine variety, then there exists an integer $n \geq 0$ such that $\Sigma_{\mathbb{P}^1}^n (X, x)$ is $\mathbb{A}^1$-contractible; in fact, $n = 2$ should suffice. □

Remark 5.5.3.12 The stronger assertion here is obtained by combining the weaker assertion and a conjecture about conservativity of $\mathbb{G}_m$-stabilization [20] (the assumption $n = 2$ essentially stems from this conservativity conjecture). Conjecture 5.5.3.11 is reminiscent of an open version of the Cannon–Edwards double suspension theorem [30, 51]. Conjecture 5.5.3.11 in conjunction with $\mathbb{A}^1$-representability of Chow groups (see Theorem 5.2.4.10) implies that if X is any topologically contractible smooth complex variety, then $CH^i(X) = 0$ for every $i > 0$. In conjunction with Theorem 5.5.2.10, Conjecture 5.5.3.11 thus implies that the generalized Serre question has a positive answer in dimension 3. We thank Tariq Syed for informing us that the same conclusion holds in dimension 4 owing to his work in [135]. □

5.5.4 $\mathbb{A}^1$-contractibility of the Koras–Russell Threefold

In what follows we will outline how Theorem 5.5.3.9 can be used to show the Russell threefold $\mathcal{KR}$ is $\mathbb{A}^1$-contractible over any base field of characteristic zero. This was carried out by Dubouloz and Fasel in [45].

Observe that $\mathcal{KR}$ contains both the affine line $\mathbb{A}^1_y$ and the affine plane $\mathbb{A}^2_{z,t}$ intersecting transversally in the origin. The idea is now to show that the inclusion $\mathbb{A}^2_{z,t} \to \mathcal{KR}$ is an $\mathbb{A}^1$-equivalence. There is a naturally induced commutative diagram of homotopy cofiber sequences

$$\begin{array}{ccccc}
\mathbb{A}^2_{z,t} \smallsetminus \{(0,0)\} & \longrightarrow & \mathbb{A}^2_{z,t} & \longrightarrow & \mathbb{P}^1_z \wedge \mathbb{P}^1_t \\
\downarrow{\scriptstyle i} & & \downarrow & & \downarrow \\
\mathcal{KR} \smallsetminus \mathbb{A}^1_y & \longrightarrow & \mathcal{KR} & \longrightarrow & \mathcal{KR}/(\mathcal{KR} \smallsetminus \mathbb{A}^1_y).
\end{array} \qquad (5.1)$$

Here the rightmost vertical map is an $\mathbb{A}^1$-equivalence induced by the inclusion $\{0\} \subset \mathbb{A}^1_y$: This follows since the normal bundle of $\mathbb{A}^1_y$ in $\mathcal{KR}$ is trivial, so that by homotopy purity 5.2.4.14 we obtain $\mathbb{A}^1$-equivalences

$$\mathcal{KR}/(\mathcal{KR} \smallsetminus \mathbb{A}^1_y) \sim_{\mathbb{A}^1} (\mathbb{A}^1_y)_+ \wedge (\mathbb{P}^1)^{\wedge 2} \sim_{\mathbb{A}^1} (\mathbb{P}^1)^{\wedge 2}.$$

By a general result we are reduced to showing that i is an $\mathbb{A}^1$-equivalence. The perhaps most technical argument in the proof consists of showing that $\mathcal{KR} \smallsetminus \mathbb{A}^1_y$ is $\mathbb{A}^1$-weak equivalent to the punctured affine space $\mathbb{A}^2_{z,t} \smallsetminus \{(0,0)\}$. We discuss this part later in this section. As a consequence the leftmost vertical map in (5.1) is a self-map of $\mathbb{A}^2_{z,t} \smallsetminus \{(0,0)\}$ up to $\mathbb{A}^1$-equivalence. As a consequence, its $\mathbb{A}^1$-homotopy class is determined by its motivic Brouwer degree (see Theorem 5.4.3.3). The computation of this degree may be turned into understanding a certain map in sheaf cohomology for the Nisnevich topology with coefficients in Milnor-Witt K-theory.

Lemma 5.5.4.1 *Let* $f : \mathbb{A}^n \smallsetminus \{0\} \to \mathbb{A}^n \smallsetminus \{0\}$ *be a morphism in* $\mathcal{H}(k)$. *Then* f *is an isomorphism if and only if*

$$f^* : H^{n-1}(\mathbb{A}^n \smallsetminus \{0\}, \mathbf{K}_n^{MW}) \to H^{n-1}(\mathbb{A}^n \smallsetminus \{0\}, \mathbf{K}_n^{MW})$$

is an isomorphism. ☐

According to Lemma 5.5.4.1 we are reduced to showing that

$$i^* : H^1(\mathcal{KR} \smallsetminus \mathbb{A}^1_y, \mathbf{K}_2^{MW}) \to H^1(\mathbb{A}^2_{z,t} \smallsetminus \{0\}, \mathbf{K}_2^{MW})$$

is an isomorphism. In effect, we consider the commutative diagram

$$
\begin{array}{ccccccc}
H^1(\mathcal{KR} \smallsetminus \mathbb{A}^1_y, \mathbf{K}_2^{MW}) & \xrightarrow{\partial} & H^2((\mathbb{P}^1)^{\wedge 2}, \mathbf{K}_2^{MW}) & \longrightarrow & H^2(\mathcal{KR}, \mathbf{K}_2^{MW}) & \longrightarrow & H^2(\mathcal{KR} \smallsetminus \mathbb{A}^1_y, \mathbf{K}_2^{MW}) \\
{\scriptstyle i^*}\downarrow & & \| & & \downarrow & & \downarrow \\
H^1(\mathbb{A}^2_{z,t} \smallsetminus \{0\}, \mathbf{K}_2^{MW}) & \xrightarrow{\partial'} & H^2((\mathbb{P}^1)^{\wedge 2}, \mathbf{K}_2^{MW}) & \longrightarrow & H^2(\mathbb{A}^2_{z,t}, \mathbf{K}_2^{MW}) & \longrightarrow & H^2(\mathbb{A}^2_{z,t} \smallsetminus \{0\}, \mathbf{K}_2^{MW})
\end{array}
$$

obtained from (5.1). One checks readily that ∂' is an isomorphism, so that i^* is an isomorphism if and only if ∂ is an isomorphism. Since ∂ is a $\mathbf{K}_0^{MW}(k)$-linear map between free $\mathbf{K}_0^{MW}(k)$-modules of rank one, it suffices to show the following assertion.

Proposition 5.5.4.2 *The connecting homomorphism* $\partial : H^1(\mathcal{KR} \smallsetminus \mathbb{A}^1_y, \mathbf{K}_2^{MW}) \to H^2((\mathbb{P}^1)^{\wedge 2}, \mathbf{K}_2^{MW})$ *is surjective.* □

Proof As the first row in the above diagram is exact, it is sufficient to prove that $H^2(\mathcal{KR}, \mathbf{K}_2^{MW}) = 0$. Fasel's projective bundle theorem [52] implies

$$
H^i(\mathcal{KR}, \mathbf{K}_j^{MW}) \cong H^{i+n}(\mathcal{KR}_+ \wedge (\mathbb{P}^1)^{\wedge n}, \mathbf{K}_{j+n}^{MW})
$$

for all $i, n \in \mathbb{N}$ and $j \in \mathbb{Z}$. Theorem 5.5.3.9 holds more generally over fields of characteristic zero and shows that $\mathcal{KR} \wedge (\mathbb{P}^1)^{\wedge n} = *$ for $n \gg 0$. Combined with the homotopy cofiber sequence

$$
(\mathbb{P}^1)^{\wedge n} \to \mathcal{KR}_+ \wedge (\mathbb{P}^1)^{\wedge n} \to \mathcal{KR} \wedge (\mathbb{P}^1)^{\wedge n}
$$

we find $H^i(\mathcal{KR}, \mathbf{K}_j^{MW}) = H^{i+n}((\mathbb{P}^1)^{\wedge n}, \mathbf{K}_{j+n}^{MW})$ for $i \geq 1$, and the latter group is trivial. □

Dubouloz and Fasel also give an alternate proof of Proposition 5.5.4.2 by means of explicit symbol calculations [45]. Moreover, the proof of $\mathbb{A}^1$-contractibility for $\mathcal{KR}$ works more generally for Koras–Russell threefolds of the first kind. It is unclear whether a similar proof works for Koras–Russell threefolds of the second kind. The main issue at stake for such a threefold X is whether there exists an $\mathbb{A}^1$-weak equivalence between the complement $X \smallsetminus \mathbb{A}^1_y$ and some punctured affine plane. On the other hand, in light of Theorem 1 from the introduction the proof is robust enough to provide many new examples of affine $\mathbb{A}^1$-contractible varieties of dimension 3.

Theorem 5.5.4.3 ([45, Corollary 1.3]) *Assume k is a field. For every integer $m \geq 2$, there exists a smooth affine morphism $\pi : X \to \mathbb{A}^{m-2}$ of relative dimension 3 whose fibers are all $\mathbb{A}^1$-contractible. Furthermore, fibers of π over k-points are pairwise non-isomorphic and stably isomorphic.* □

Question 5.5.4.4 Looking further forward: can one characterize affine spaces among affine $\mathbb{A}^1$-contractible varieties in a motivic version of the Poincaré conjecture (e.g., by defining some notion of $\mathbb{A}^1$-fundamental group at infinity)? □

5.5.5   Koras–Russell Fiber Bundles

The geometry becomes much more pronounced in the proof showing that $\mathcal{KR} \setminus \mathbb{A}_y^1$ is $\mathbb{A}^1$-weak equivalent to the punctured affine space $\mathbb{A}_{z,t}^2 \setminus \{(0,0)\}$; the salient geometric features of Koras–Russell threefolds of the first kind have been further developed into the context of Koras–Russell fiber bundles introduced in [47]. In the following we assume k is an algebraically closed field of characteristic zero.

Definition 5.5.5.1 Suppose $s(x) \in k[x]$ has positive degree and let $R(x, y, t) \in k[x, y, t]$. Define the closed subscheme $X(s, R)$ of $\mathbb{A}_x^1 \times \mathbb{A}^3 = \mathrm{Spec}(k[x][y, z, t])$ by the equation

$$\{s(x)z = R(x, y, t)\}.$$

We say that the projection map

$$\rho := \mathrm{pr}_x : X(s, R) \to \mathbb{A}_x^1 \tag{5.1}$$

defines a *Koras–Russell fiber bundle* if

(a) $X(s, R)$ is a smooth scheme, and
(b) For every zero x_0 of $s(x)$, the zero locus in $\mathbb{A}^2 = \mathrm{Spec}(k[y, t])$ of the polynomial $R(x_0, y, t)$ is an integral rational plane curve with a unique place at infinity and at most unibranch singularities. □

Remark 5.5.5.2 One can show that a Koras–Russell fiber bundle is isomorphic to $\mathbb{A}^3$ if and only if for every zero x_0 of $s(x)$ the curve $\{R(x_0, y, t) = 0\}$ is isomorphic to $\mathbb{A}^1$. □

For concreteness we discuss two classes of examples of Koras–Russell fiber bundles.

Example 5.5.5.3 Deformed Koras–Russell threefolds of the first kind are defined as

$$X(n, \alpha_i, p) := \{x^n z = y^{\alpha_1} + t^{\alpha_2} + xp(x, y, t)\}, \tag{5.2}$$

where $n, \alpha_i \geq 2$ are integers, α_1 and α_2 are coprime, and $p(x, y, t) \in k[x, y, t]$ satisfies $p(0, 0, 0) \in k^*$. By Dubouloz and Fasel [45] it is known that $X(n, \alpha_i, p)$ is $\mathbb{A}^1$-contractible when $p(x, y, t) = q(x) \in k[x]$ and $q(0) \in k^*$. Note that $X(n, \alpha_i, p)$ is smooth according to the Jacobian criterion since $p(0, 0, 0) \in k^*$. Moreover, the unique singular fiber of the projection map

$$\mathrm{pr}_x : X(n, \alpha_i, p) \to \mathbb{A}_x^1 \tag{5.3}$$

is a cylinder on the cuspidal curve $\Gamma_{\alpha_1, \alpha_2} := \{y^{\alpha_1} + t^{\alpha_2} = 0\} \subset \mathbb{A}^2$, which is $\mathbb{A}^1$-contractible

$$\mathrm{pr}_x^{-1}(0) = \Gamma_{\alpha_1, \alpha_2} \times \mathbb{A}_z^1 \sim_{\mathbb{A}^1} *. \tag{5.4}$$

Here (5.3) is a flat $\mathbb{A}^2$-fibration restricting to a trivial $\mathbb{A}^2$-bundle over $\mathbb{A}^1_x \smallsetminus \{0\}$ and $X(n, \alpha_i, p)$ is factorial. The $\mathbb{A}^1$-homotopy theory of deformed Koras–Russell threefolds of the first kind (5.2) is essentially governed by (5.3) and (5.4).   □

Example 5.5.5.4 For $1 \le i \le m$, choose distinct linear forms $l_i(x) = (x - x_i) \in k[x]$, $n_i, \alpha_i, \beta_i \ge 2$, where α_i and β_i are coprime, and $a \in k^\times$. We define $X_m(n_i, \alpha_i, \beta_i, a)$ or simply X_m by the equation

$$X_m = X_m(n_i, \alpha_j, \beta_j, a) := \left\{ \left(\prod_{i=1}^m l_i(x)^{n_i} \right) z = \sum_{i=1}^m \left(\left(\prod_{j \ne i} l_j(x) \right) (y^{\alpha_i} + t^{\beta_i}) \right) + a \prod_{i=1}^m l_i(x) \right\}.$$

(5.5)

In the case of two degenerate fibers, (5.5) takes the form

$$X_2 = \{(x - x_1)^{n_1}(x - x_2)^{n_2} z = (x - x_1)(y^{\alpha_2} + t^{\beta_2}) + (x - x_2)(y^{\alpha_1} + t^{\beta_1}) + a(x - x_1)(x - x_2)\}.$$

(5.6)

For all m the Makar-Limanov invariant of $X_m(n_i, \alpha_j, \beta_j, a)$ equals $k[x]$; hence it is non-isomorphic to $\mathbb{A}^3$. Moreover, the projection map

$$\mathrm{pr}_x : X_m(n_i, \alpha_j, \beta_j, a) \to \mathbb{A}^1_x \qquad (5.7)$$

defines a trivial $\mathbb{A}^2$-bundle over the punctured affine line $\mathbb{A}^1_x \smallsetminus \{x_1, \ldots, x_m\}$. Its fiber over the closed point $x_i \in \mathbb{A}^1_x$ is isomorphic to the cylinder on the cuspidal curve $\Gamma_{\alpha_i, \beta_i} : \{y^{\alpha_i} + t^{\beta_i} = 0\}$. By counting closed fibers non-isomorphic to $\mathbb{A}^2$ one concludes that X_m and $X_{m'}$ are non-isomorphic when $m \ne m'$.   □

Now we turn to the geometric properties of deformed Koras–Russell threefolds as in Example 5.5.5.3. In particular, this will explain the $\mathbb{A}^1$-equivalence between $\mathcal{KR} \smallsetminus \mathbb{A}^1_y$ and the punctured affine plane.

There is an induced $\mathbb{G}_a$-action on $X(n, \alpha_i, p)$ determined by the locally nilpotent derivation

$$\partial = x^n \frac{\partial}{\partial y} + (\alpha_1 y^{\alpha_1 - 1} + x \frac{\partial}{\partial y} p(x, y, t)) \frac{\partial}{\partial z},$$

on the coordinate ring of $X(n, \alpha_i, p)$, with fixed point locus the affine line $\{x = y = t = 0\} \cong \mathbb{A}^1_z$. The geometric quotient $X(n, \alpha_i, p) \to X(n, \alpha_i, p)/\mathbb{G}_a$ yields an $\mathbb{A}^1$-bundle $X(n, \alpha_i, p) \smallsetminus \mathbb{A}^1_z \to \mathfrak{S}(\alpha_1, \alpha_2)$ in the category of algebraic spaces [88]. In fact there exists a factorization

$$X(n, \alpha_i, p) \smallsetminus \mathbb{A}^1_z \xrightarrow{\;\;\pi_1\;\;} \mathbb{A}^2_{x,t} \smallsetminus \{(0,0)\}$$
$$\rho \searrow \qquad \swarrow \delta$$
$$\mathfrak{S}(\alpha_1, \alpha_2),$$

(5.8)

where $\pi_|$ is the restriction of $\pi = \mathrm{pr}_{x,t} : \mathcal{X}(n, \alpha_i, p) \to \mathbb{A}^2_{x,t}$ to $\mathcal{X}(n, \alpha_i, p) \smallsetminus \mathbb{A}^1_z$. To construct (5.8) we form a cyclic Galois cover of $\mathbb{A}^2_{x,t}$ of order α_1 and hence of $\mathcal{X}(n, \alpha_i, p)$ by pullback via π. The maps arise as geometric quotients for μ_{α_1}-equivariant maps by gluing copies of $\mathbb{A}^2 \smallsetminus \{(0,0)\}$ via a family of cuspidal curves. Here ρ is an étale locally trivial $\mathbb{A}^1$-bundle, we have $\mathfrak{S}(\alpha_1, \alpha_2) \cong \mathfrak{S}(\alpha_1, 1)$, and both of the projection maps for the smooth quasi-affine 4-fold

$$(\mathcal{X}(n, \alpha_i, p) \smallsetminus \mathbb{A}^1_z) \times_{\mathfrak{S}(\alpha_1, \alpha_2)} (\mathcal{X}(n, \alpha_1, 1, p) \smallsetminus \mathbb{A}^1_z) \tag{5.9}$$

are Zariski locally trivial $\mathbb{A}^1$-bundles, and hence $\mathbb{A}^1$-weak equivalences. The fiber product in (5.9) is formed in algebraic spaces over the punctured affine plane $\mathbb{A}^2_{x,t} \smallsetminus \{(0,0)\}$. Furthermore, the projection map $\mathrm{pr}_x : \mathcal{X}(n, \alpha_1, 1, p) \to \mathbb{A}^1_x$ is a trivial $\mathbb{A}^2$-bundle. It follows that $\mathcal{X}(n, \alpha_1, 1, p) \cong \mathbb{A}^3_{x,y,z}$ and we can finally conclude that there exist $\mathbb{A}^1$-weak equivalences

$$\mathcal{X}(n, \alpha_i, p) \smallsetminus \mathbb{A}^1_z \sim_{\mathbb{A}^1} \mathcal{X}(n, \alpha_1, 1, p) \smallsetminus \mathbb{A}^1_z \sim_{\mathbb{A}^1} \mathbb{A}^2_{x,t} \smallsetminus \{(0,0)\}.$$

The currently most general result concerning $\mathbb{A}^1$-contractibility of Koras–Russell fiber bundles was shown in [47].

Theorem 5.5.5.5 *Suppose $\rho : \mathcal{X}(s, R) \to \mathbb{A}^1_x$ is a Koras–Russell fiber bundle with basepoint the origin. The S^1-suspension and hence the $\mathbb{P}^1$-suspension of $\mathcal{X}(s, R)$ are $\mathbb{A}^1$-contractible:*

$$\mathcal{X}(s, R) \wedge S^1 \sim_{\mathbb{A}^1} \mathcal{X}(s, R) \wedge \mathbb{P}^1 \sim_{\mathbb{A}^1} *.$$

Question 5.5.5.6 Can one generalize the notion of a Koras–Russell fiber bundle and show Theorem 5.5.5.5 over arbitrary fields of characteristic zero?   □

Remark 5.5.5.7 Work on Question 5.5.5.6 is likely to involve base change arguments to an algebraic closure.   □

Acknowledgments Asok was partially supported by National Science Foundation Awards DMS-1254892 and DMS-1802060. Østvær was partially supported by RCN Frontier Research Group Project no. 250399, Friedrich Wilhelm Bessel Research Award from the Humboldt Foundation, and Nelder Visiting Fellowship from Imperial College London. The authors are grateful to Ben Antieau, Brent Doran, Adrien Dubouloz, Jean Fasel, Marc Hoyois, Amalendu Krishna, Fabien Morel, Sabrina Pauli, and Ben Williams for collaborative efforts and many interesting discussions on the topics surveyed in this paper. The authors would also like to thank W. Danielewski for interesting comments and corrections to an earlier version of this paper.

References

1. S. S. Abhyankar, W. Heinzer, P. Eakin, On the uniqueness of the coefficient ring in a polynomial ring. J. Algebra **23**, 310–342 (1972)
2. R. Achet, The Picard group of the forms of the affine line and of the additive group. J. Pure Appl. Algebra **221**(11), 2838–2860 (2017). https://doi.org/10.1016/j.jpaa.2017.02.003
3. J.F. Adams, *Stable Homotopy and Generalised Homology* (University of Chicago, Chicago, Ill.-London, 1974). Chicago Lectures in Mathematics
4. A. Ananyevskiy, A. Luzgarev, *Private Communication* (2016)
5. B. Antieau, E. Elmanto, A primer for unstable motivic homotopy theory, in *Surveys on Recent Developments in Algebraic Geometry.* Proceedings Symposium of Pure Mathematical , vol. 95 (American Mathematical Society, Providence, RI, 2017), pp. 305–370
6. T. Asanuma, Non-linearizable algebraic k^*-actions on affine spaces. Invent. Math. **138**(2), 281–306 (1999)
7. T. Asanuma, N. Gupta, On 2-stably isomorphic four-dimensional affine domains. J. Commut. Algebra **10**(2), 153–162 (2018)
8. A. Asok, Motives of some acyclic varieties. Homology Homotopy Appl. **13**(2), 329–335 (2011)
9. A. Asok, Birational invariants and $\mathbb{A}^1$-connectedness. J. Reine Angew. Math. **681**, 39–64 (2013)
10. A. Asok, B. Doran, On unipotent quotients and some $\mathbb{A}^1$-contractible smooth schemes. Int. Math. Res. Pap. IMRP **2**, Art. ID rpm005, 51 (2007)
11. A. Asok, B. Doran, Vector bundles on contractible smooth schemes. Duke Math. J. **143**(3), 513–530 (2008)
12. A. Asok, J. Fasel, A cohomological classification of vector bundles on smooth affine threefolds. Duke Math. J. **163**, 2561–2601 (2014)
13. A. Asok, J. Fasel, Splitting vector bundles outside the stable range and $\mathbb{A}^1$-homotopy sheaves of punctured affine space. J. Am. Math. Soc. **28**, 1031–1062 (2015)
14. A. Asok, C. Haesemeyer, Stable $\mathbb{A}^1$-homotopy and R-equivalence. J. Pure Appl. Algebra **215**(10), 2469–2472 (2011)
15. A. Asok, F. Morel, Smooth varieties up to $\mathbb{A}^1$-homotopy and algebraic h-cobordisms. Adv. Math. **227**(5), 1990–2058 (2011)
16. A. Asok, B. Doran, J. Fasel, Smooth models of motivic spheres and the clutching construction. Int. Math. Res. Not. IMRN **6**, 1890–1925 (2017)
17. A. Asok, M. Hoyois, M. Wendt, Affine representability results in $\mathbb{A}^1$-homotopy theory I: vector bundles. Duke Math. J. **166**(10), 1923–1953 (2017)
18. M.F. Atiyah, G.B. Segal, Equivariant K-theory and completion. J. Differ. Geom. **3**, 1–18 (1969)
19. J. Ayoub, Motives and algebraic cycles: a selection of conjectures and open questions, in *Hodge Theory and L^2-analysis.* Advanced Lecture Mathematics (ALM), vol. 39 (International Press, Somerville, MA, 2017), pp. 87–125
20. T. Bachmann, M. Yakerson, *Towards conservativity of $\mathbb{G}_m$-stabilization* (2018). https://arxiv.org/abs/1811.01541
21. C. Balwe, A. Sawant, R-equivalence and $\mathbb{A}^1$-connectedness in anisotropic groups. Int. Math. Res. Not. (IMRN) **2015**(22), 11816–11827 (2015)
22. C. Balwe, A. Hogadi, A. Sawant, $\mathbb{A}^1$-connected components of schemes. Adv. Math. **282**, 335–361 (2015)
23. R. Barlow, A simply connected surface of general type with $p_g = 0$. Invent. Math. **79**(2), 293–301 (1985)
24. A. Beauville, J.-L. Colliot-Thélène, J.-J. Sansuc, P. Swinnerton-Dyer, Variétés stablement rationnelles non rationnelles. Ann. Math. (2) **121**(2), 283–318 (1985)
25. J.P. Bell, The equivariant Grothendieck groups of the Russell-Koras threefolds. Canad. J. Math. **53**(1), 3–32 (2001)

26. B.A. Blander, Local projective model structures on simplicial presheaves. K-Theory **24**(3), 283–301 (2001)

27. S. Bloch, Algebraic cycles and higher K-theory. Adv. Math. **61**(3), 267–304 (1986)

28. S. Bloch, The moving lemma for higher Chow groups. J. Algebraic Geom. **3**(3), 537–568 (1994)

29. M. Brion, Some structure theorems for algebraic groups, in *Algebraic Groups: Structure and Actions. Proceedings Symposium of Pure Mathematical* , vol. 94 (American Mathematical Society, Providence, RI, 2017), pp. 53–126

30. J.W. Cannon, Shrinking cell-like decompositions of manifolds. Codimension three. Ann. Math. (2) **110**(1), 83–112 (1979)

31. J. Carlson, ed. *The Poincaré conjecture. Clay Mathematics Proceedings*, vol. 19. (American Mathematical Society/Clay Mathematics Institute, Providence/Cambridge, 2014). Selected lectures from the Clay Research Conference on the Resolution of the Poincaré Conjecture held at the Institute Henri Poincaré, Paris, June 8–9, 2010

32. F. Charles, Conjugate varieties with distinct real cohomology algebras. J. Reine Angew. Math. **630**, 125–139 (2009)

33. D.B. Coleman, E.E. Enochs, Isomorphic polynomial rings. Proc. Am. Math. Soc. **27**, 247–252 (1971)

34. J.-L. Colliot-Thélène, J.-J. Sansuc, The rationality problem for fields of invariants under linear algebraic groups (with special regards to the Brauer group), in *Algebraic Groups and Homogeneous Spaces*. Tata Institue of Fundamental Research Studia Mathematica, vol. 19 (Tata Institue of Fundamental Research, Mumbai, 2007), pp. 113–186

35. M.L. Curtis, K.W. Kwun, Infinite sums of manifolds. Topology **3**, 31–42 (1965)

36. W. Danielewski, On a cancellation problem and automorphism group of affine algebraic varieties. Preprint (1988)

37. W. Danielewski, *Private Communication* (2019)

38. P. Deligne, Voevodsky's lectures on motivic cohomology 2000/2001, in *Algebraic Topology. Abel Symposium*, vol. 4 (Springer, Berlin, 2009), pp. 355–409

39. J.-P. Demailly, Hyperbolic algebraic varieties and holomorphic differential equations. Acta Math. Vietnam. **37**(4), 441–512 (2012)

40. A. Dubouloz, Danielewski-Fieseler surfaces. Transform. Groups **10**(2), 139–162 (2005)

41. A. Dubouloz, Additive group actions on Danielewski varieties and the cancellation problem. Math. Z. **255**(1), 77–93 (2007)

42. A. Dubouloz, The cylinder over the Koras-Russell cubic threefold has a trivial Makar-Limanov invariant. Transform. Groups **14**(3), 531–539 (2009)

43. A. Dubouloz, Affine open subsets in $\mathbb{A}^3$ without the cancellation property, in *Commutative Algebra and Algebraic Geometry (CAAG-2010)*. Ramanujan Mathematical Society Lecture Notes Series, vol. 17 (Ramanujan Mathematical Society, Mysore, 2013), pp. 63–67

44. A. Dubouloz, Affine surfaces with isomorphic $\mathbb{A}^2$-cylinders. Kyoto J. Math. **59**(1), 181–193 (2019). Advance publication

45. A. Dubouloz, J. Fasel, Families of $\mathbb{A}^1$-contractible affine threefolds. Algebr. Geom. **5**(1), 1–14 (2018)

46. A. Dubouloz, L. Moser-Jauslin, P.-M. Poloni, Noncancellation for contractible affine three-folds. Proc. Am. Math. Soc. **139**(12), 4273–4284 (2011)

47. A. Dubouloz, S. Pauli, P.A. Østvær, $\mathbb{A}^1$-contractibility of affine modifications. Internat. J. Math. **30**(14), 1950069, 34 (2019)

48. D. Dugger, Universal homotopy theories. Adv. Math. **164**(1), 144–176 (2001)

49. D. Dugger, D.C. Isaksen, Topological hypercovers and $\mathbb{A}^1$-realizations. Math. Z. **246**(4), 667–689 (2004)

50. B.I. Dundas, O. Röndigs, P.A. Østvær, Motivic functors. Doc. Math. **8**, 489–525 (electronic) (2003)

51. R.D. Edwards, *Suspensions of Homology Spheres* (2006). https://arxiv.org/abs/math/0610573

52. J. Fasel, The projective bundle theorem for I^j-cohomology. J. K-Theory **11**(2), 413–464 (2013)

53. T. Fujita, On Zariski problem, Proc. Jpn. Acad. Ser. A Math. Sci. **55**(3), 106–110 (1979)
54. T. Fujita, On the topology of noncomplete algebraic surfaces. J. Fac. Sci. Univ. Tokyo Sect. IA Math. **29**(3), 503–566 (1982)
55. W. Fulton, Intersection theory, in *Ergebnisse der Mathematik und ihrer Grenzgebiete. 3. Folge. A Series of Modern Surveys in Mathematics*, vol. 2, 2nd edn. (Springer, Berlin, 1998)
56. D. Gepner, V. Snaith, On the motivic spectra representing algebraic cobordism and algebraic K-theory. Doc. Math. **14**, 359–396 (2009)
57. L.C. Glaser, Uncountably many contractible open 4-manifolds. Topology **6**, 37–42 (1967)
58. J. Glimm, Two Cartesian products which are Euclidean spaces. Bull. Soc. Math. France **88**, 131–135 (1960)
59. N. Gupta, On the cancellation problem for the affine space A^3 in characteristic p. Invent. Math. **195**(1), 279–288 (2014)
60. N. Gupta, On Zariski's cancellation problem in positive characteristic. Adv. Math. **264**, 296–307 (2014)
61. N. Gupta, On the family of affine threefolds $x^m y = F(x, z, t)$. Compos. Math. **150**(6), 979–998 (2014)
62. N. Gupta, A survey on Zariski cancellation problem. Indian J. Pure Appl. Math. **46**(6), 865–877 (2015)
63. R. V. Gurjar, Affine varieties dominated by C^2. Comment. Math. Helv. **55**(3), 378–389 (1980). https://doi.org/10.1007/BF02566694
64. R.V. Gurjar, A.R. Shastri, On the rationality of complex homology 2-cells. I. J. Math. Soc. Jpn. **41**(1), 37–56 (1989)
65. R.V. Gurjar, A.R. Shastri, On the rationality of complex homology 2-cells. II. J. Math. Soc. Jpn. **41**(2), 175–212 (1989)
66. J.J. Gutiérrez, O. Röndigs, M. Spitzweck, P.A. Østvær, Motivic slices and coloured operads. J. Topol. **5**(3), 727–755 (2012)
67. M. Hazewinkel, C.F. Martin, A short elementary proof of Grothendieck's theorem on algebraic vectorbundles over the projective line. J. Pure Appl. Algebra **25**(2), 207–211 (1982)
68. J. Heller, A. Krishna, P. A. Østvær, Motivic homotopy theory of group scheme actions. J. Topol. **8**(4), 1202–1236 (2015)
69. J. Heller, M. Voineagu, P.A. Østvær, Topological comparison theorems for Bredon motivic cohomology. Trans. Am. Math. Soc. **371**(4), 2875–2921 (2019)
70. M. Hochster, Nonuniqueness of coefficient rings in a polynomial ring. Proc. Am. Math. Soc. **34**, 81–82 (1972)
71. A. Holschbach, J. Schmidt, J. Stix, Étale contractible varieties in positive characteristic. Algebra Number Theory **8**(4), 1037–1044 (2014)
72. M. Hovey, Model categories, in *Mathematical Surveys and Monographs*, vol. 63 (American Mathematical Society, Providence, RI, 1999)
73. M. Hovey, Spectra and symmetric spectra in general model categories. J. Pure Appl. Algebra **165**(1), 63–127 (2001)
74. M. Hoyois, A. Krishna, P.A. Østvær, A^1-contractibility of Koras-Russell threefolds. Algebr. Geom. **3**(4), 407–423 (2016)
75. A. Huber, Slice filtration on motives and the Hodge conjecture (with an appendix by J. Ayoub). Math. Nachr. **281**(12), 1764–1776 (2008)
76. B. Hughes, A. Ranicki, Ends of complexes, in *Cambridge Tracts in Mathematics*, vol. 123 (Cambridge University, Cambridge, 1996)
77. S. Iitaka, T. Fujita, Cancellation theorem for algebraic varieties. J. Fac. Sci. Univ. Tokyo Sect. IA Math. **24**(1), 123–127 (1977)
78. D.C. Isaksen, Etale realization on the A^1-homotopy theory of schemes. Adv. Math. **184**(1), 37–63 (2004)
79. D. Isaksen, P.A. Østvær, *Motivic Stable Homotopy Groups* (2018). https://arxiv.org/abs/1811.05729
80. J.F. Jardine, Simplicial presheaves. J. Pure Appl. Algebra **47**(1), 35–87 (1987)
81. J.F. Jardine, Motivic symmetric spectra. Doc. Math. **5**, 445–553 (electronic) (2000)

82. J.F. Jardine, Representability theorems for presheaves of spectra. J. Pure Appl. Algebra **215**(1), 77–88 (2011)

83. J.F. Jardine, Local homotopy theory, in *Springer Monographs in Mathematics* (Springer, New York, 2015)

84. Z. Jelonek, On the cancellation problem. Math. Ann. **344**(4), 769–778 (2009)

85. J.P. Jouanolou, *Une suite exacte de Mayer-Vietoris en K-théorie algébrique*. Lecture Notes in Mathematical, vol. 341 (1973), pp. 293–316

86. Sh. Kaliman, Polynomials with general $\mathbf{C}^2$-fibers are variables. Pacific J. Math. **203**(1), 161–190 (2002)

87. J.M. Kister, D.R. McMillan, Jr, Locally euclidean factors of E^4 which cannot be imbedded in E^3. Ann. Math. (2) **76**, 541–546 (1962)

88. D. Knutson, *Algebraic Spaces*. Lecture Notes in Mathematics, vol. 203 (Springer, Berlin, 1971)

89. M. Koras, P. Russell, Contractible threefolds and $\mathbf{C}^*$-actions on $\mathbf{C}^3$. J. Algebraic Geom. **6**(4), 671–695 (1997)

90. H. Kraft, Challenging problems on affine n-space. Astérisque, (237):Exp. No. 802 **5**, 295–317 (1996). Séminaire Bourbaki, vol. 1994/95

91. S. Kumar, A generalization of the Conner conjecture and topology of Stein spaces dominated by $\mathbf{C}^n$. Topology **25**(4), 483–493 (1986)

92. T.Y. Lam, *Serre's problem on projective modules*. Springer Monographs in Mathematics (Springer, Berlin, 2006)

93. M. Levine, Convergence of Voevodsky's slice tower. Doc. Math. **18**, 907–941 (2013)

94. M. Levine, A comparison of motivic and classical stable homotopy theories. J. Topol. **7**(2), 327–362 (2014)

95. M. Levine, An overview of motivic homotopy theory. Acta Math. Vietnam. **41**(3), 379–407 (2016)

96. H. Lindel, On the Bass-Quillen conjecture concerning projective modules over polynomial rings. Invent. Math. **65**(2), 319–323 (1981/1982)

97. S.S.Y. Lu, D.-Q. Zhang, Positivity criteria for log canonical divisors and hyperbolicity. J. Reine Angew. Math. **726**, 173–186 (2017)

98. L. Makar-Limanov, On the hypersurface $x + x^2 y + z^2 + t^3 = 0$ in $\mathbb{C}^4$ or a $\mathbb{C}^3$-like threefold which is not $\mathbb{C}^3$. Isr. J. Math. **96**, 419–429 (1996)

99. B. Mazur, A note on some contractible 4-manifolds. Ann. Math. (2) **73**, 221–228 (1961)

100. D.R. McMillan, Jr., Some contractible open 3-manifolds. Trans. Am. Math. Soc. **102**, 373–382 (1962)

101. J. Milnor, Algebraic K-theory and quadratic forms. Invent. Math. **9**, 318–344 (1969/1970)

102. J. Milnor, The work of M. H. Freedman [MR0934211 (89d:01064)], in *Fields Medallists' Lectures*. World Science of Series 20th Century Mathematical, vol. 5 (World Science Publication, River Edge, 1997), pp. 405–408

103. J. Milnor, D. Husemoller, *Symmetric Bilinear Forms* (Springer, New York, 1973). Ergebnisse der Mathematik und ihrer Grenzgebiete, Band 73

104. M. Miyanishi, T. Sugie, Affine surfaces containing cylinderlike open sets. J. Math. Kyoto Univ. **20**(1), 11–42 (1980)

105. M. Miyanishi, S. Tsunoda, Absence of the affine lines on the homology planes of general type. J. Math. Kyoto Univ. **32**(3), 443–450 (1992)

106. N. Mohan Kumar, M.P. Murthy, Algebraic cycles and vector bundles over affine three-folds. Ann. Math. (2) **116**(3), 579–591 (1982)

107. F. Morel, An introduction to $\mathbb{A}^1$-homotopy theory, in *Contemporary Developments in Algebraic K-theory*. ICTP Lecture Notes, vol. XV (electronic) (Abdus Salam International Central Theoretical Physics, Trieste, 2004), pp. 357–441

108. F. Morel, The stable $\mathbb{A}^1$-connectivity theorems. K-Theory **35**(1–2), 1–68 (2005)

109. F. Morel, $\mathbb{A}^1$-algebraic topology, in *International Congress of Mathematicians*, vol. II (European Mathematical Society, Zürich, 2006), pp. 1035–1059

110. F. Morel, $\mathbb{A}^1$-algebraic topology over a field, in *Lecture Notes in Mathematics*, vol. 2052 (Springer, Heidelberg, 2012)
111. F. Morel, V. Voevodsky, $\mathbb{A}^1$-homotopy theory of schemes. Inst. Hautes Études Sci. Publ. Math. **1999**(90), 45–143 (2001)
112. M.P. Murthy, Cancellation problem for projective modules over certain affine algebras, in *Algebra, Arithmetic and Geometry, Part I, II (Mumbai, 2000)*. Tata Institue of Fundamental Research Studia Mathematica, vol. 16 (Tata Institue of Fundamental Research, Bombay, 2002), pp. 493–507
113. N. Naumann, M. Spitzweck, Brown representability in $\mathbb{A}^1$-homotopy theory. J. K-Theory **7**(3), 527–539 (2011)
114. I. Panin, C. Walter, *Quaternionic Grassmannians and Pontryagin Classes in Algebraic Geometry* (2010). Preprint http://arxiv.org/abs/1011.0649
115. V. Poenaru, Les decompositions de l'hypercube en produit topologique. Bull. Soc. Math. France **88**, 113–129 (1960)
116. D. Quillen, Higher algebraic K-theory. I, in *Algebraic K-theory, I: Higher K-theories (Proceedings of the Conference, Battelle Memorial Institute, Seattle, Washington, 1972)*. Lecture Notes in Mathematical, vol. 341 (Springer, Berlin, 1973), pp. 85–147
117. D. Quillen, Projective modules over polynomial rings. Invent. Math. **36**, 167–171 (1976)
118. C.P. Ramanujam, A topological characterisation of the affine plane as an algebraic variety. Ann. Math. (2) **94**, 69–88 (1971)
119. J. Riou, Dualité de Spanier-Whitehead en géométrie algébrique. C. R. Math. Acad. Sci. Paris **340**(6), 431–436 (2005)
120. A.A. Rojtman, The torsion of the group of 0-cycles modulo rational equivalence. Ann. Math. (2) **111**(3), 553–569 (1980)
121. P. Russell, On affine-ruled rational surfaces. Math. Ann. **255**(3), 287–302 (1981)
122. O. Röndigs, P.A. Østvær, Modules over motivic cohomology. Adv. Math. **219**(2), 689–727 (2008)
123. O. Röndigs, M. Spitzweck, P.A. Østvær, The first stable homotopy groups of motivic spheres. Ann. Math. (2) **189**(1), 1–74 (2019)
124. P. Russell, Forms of the affine line and its additive group. Pacific J. Math. **32**, 527–539 (1970). http://projecteuclid.org/euclid.pjm/1102977378
125. P. Russell, Cancellation, in *Automorphisms in Birational and Affine Geometry*. Springer Proceedings of Mathematical Statistical, vol. 79 (Springer, Cham, 2014), pp. 495–518
126. M. Schlichting, Euler class groups and the homology of elementary and special linear groups. Adv. Math. **320**, 1–81 (2017)
127. M. Schlichting, G.S. Tripathi, Geometric models for higher Grothendieck-Witt groups in $\mathbb{A}^1$-homotopy theory. Math. Ann. **362**(3–4), 1143–1167 (2015)
128. B. Segre, Sur un problème de M. Zariski, in *Algèbre et Théorie des Nombres*. Colloques Internationaux du Centre National de la Recherche Scientifique, vol. 24 (Centre National de la Recherche Scientifique, Paris, 1950), pp. 135–138
129. J.-P. Serre, Faisceaux algébriques cohérents. Ann. Math. (2) **61**, 197–278 (1955)
130. L.C. Siebenmann, On detecting Euclidean space homotopically among topological manifolds. Invent. Math. **6**, 245–261 (1968)
131. M. Spitzweck, P.A. Østvær, The Bott inverted infinite projective space is homotopy algebraic K-theory. Bull. Lond. Math. Soc. **41**(2), 281–292 (2009)
132. J. Stallings, The piecewise-linear structure of Euclidean space. Proc. Cambridge Philos. Soc. **58**, 481–488 (1962)
133. A.A. Suslin, Projective modules over polynomial rings are free. Dokl. Akad. Nauk SSSR **229**(5), 1063–1066 (1976)
134. A. Suslin, V. Voevodsky, Singular homology of abstract algebraic varieties. Invent. Math. **123**(1), 61–94 (1996)
135. T. Syed, *The Cancellation of Projective Modules of Rank 2 with a Trivial Determinant* (2019). https://arxiv.org/abs/1902.01130

136. *Théorie des Intersections et théorème de Riemann-Roch.* Lecture Notes in Mathematics, vol. 225 (Springer, Berlin, 1971). Séminaire de Géométrie Algébrique du Bois-Marie 1966–1967 (SGA 6), Dirigé par P. Berthelot, A. Grothendieck et L. Illusie. Avec la collaboration de D. Ferrand, J. P. Jouanolou, O. Jussila, S. Kleiman, M. Raynaud et J. P. Serre.
137. Z. Tian, H.R. Zong, One-cycles on rationally connected varieties. Compos. Math. **150**(3), 396–408 (2014)
138. T. tom Dieck, T. Petrie, Contractible affine surfaces of Kodaira dimension one. Jpn. J. Math. (N.S.) **16**(1), 147–169 (1990)
139. C. Traverso, Seminormality and Picard group. Ann. Scuola Norm. Sup. Pisa (3) **24**, 585–595 (1970)
140. A. van de Ven, Analytic compactifications of complex homology cells. Math. Ann. **147**, 189–204 (1962)
141. V. Voevodsky, $\mathbb{A}^1$-homotopy theory, in *Proceedings of the International Congress of Mathematicians (Berlin, 1998)*, vol. I (1998), pp. 579–604, number Extra Vol. I (electronic)
142. V. Voevodsky, Motivic cohomology groups are isomorphic to higher Chow groups in any characteristic. Int. Math. Res. Not. **2002**(7), 351–355 (2002)
143. V. Voevodsky, Open problems in the motivic stable homotopy theory. I, in *Motives, Polylogarithms and Hodge Theory, Part I (Irvine, CA, 1998)*. International Press Lecture Series, vol. 3 (International Press, Somerville, MA, 2002), pp. 3–34
144. V. Voevodsky, Reduced power operations in motivic cohomology. Publ. Math. Inst. Hautes Études Sci. **98**, 1–57 (2003)
145. V. Voevodsky, Motivic cohomology with $\mathbf{Z}/2$-coefficients. Publ. Math. Inst. Hautes Études Sci. **2003**(98), 59–104 (2003)
146. V. Voevodsky, Unstable motivic homotopy categories in Nisnevich and CDH-topologies. J. Pure Appl. Algebra **214**(8), 1399–1406 (2010)
147. F. Waldhausen, Algebraic K-theory of spaces. in *Algebraic and Geometric Topology (New Brunswick, N.J., 1983)*. Lecture Notes in Mathematical, vol. 1126 (Springer, Berlin, 1985), pp. 318–419
148. C.A. Weibel, Homotopy algebraic K-theory, in *Algebraic K-theory and Algebraic Number Theory (Honolulu, HI, 1987)*. Contemporary Mathematics, vol. 83 (American Mathematical Society, Providence, RI, 1989), pp. 461–488
149. J.H.C. Whitehead, Certain theorems about three-dimensional manifolds. I. Q. J. Math., Oxf. Ser. **5**, 308–320 (1934)
150. J.H.C. Whitehead, A certain open manifold whose group is unity. Q. J. Math., Oxf. Ser. **6**, 268–279 (1935)
151. K. Wickelgren, B. Williams, *Unstable Motivic Homotopy Theory* (2019). https://arxiv.org/abs/1902.08857
152. J. Wildeshaus, The boundary motive: definition and basic properties. Compos. Math. **142**(3), 631–656 (2006)
153. J. Winkelmann, On free holomorphic $\mathbf{C}$-actions on $\mathbf{C}^n$ and homogeneous Stein manifolds. Math. Ann. **286**(1–3), 593–612 (1990)
154. M.G. Zaĭdenberg, Isotrivial families of curves on affine surfaces, and the characterization of the affine plane. Izv. Akad. Nauk SSSR Ser. Mat. **51**(3), 534–567, 688 (1987)
155. M.G. Zaĭdenberg, Additions and corrections to the paper: "Isotrivial families of curves on affine surfaces, and the characterization of the affine plane" [Izv. Akad. Nauk SSSR Ser. Mat. **51**(3), 534–567 (1987); MR0903623 (88k:14021)]. Izv. Akad. Nauk SSSR Ser. Mat., **55**(2), 444–446 (1991)
156. M. Zaĭdenberg, Exotic algebraic structures on affine spaces. Algebra i Analiz **11**(5), 3–73 (1999)

Index

© The Author(s), under exclusive license to Springer Nature Switzerland AG 2021   213
F. Neumann, A. Pál (eds.), *Homotopy Theory and Arithmetic Geometry – Motivic and Diophantine Aspects*, Lecture Notes in Mathematics 2292,
https://doi.org/10.1007/978-3-030-78977-0

LECTURE NOTES IN MATHEMATICS  Springer

Editors in Chief: J.-M. Morel, B. Teissier;

Editorial Policy

1. Lecture Notes aim to report new developments in all areas of mathematics and their applications – quickly, informally and at a high level. Mathematical texts analysing new developments in modelling and numerical simulation are welcome.

   Manuscripts should be reasonably self-contained and rounded off. Thus they may, and often will, present not only results of the author but also related work by other people. They may be based on specialised lecture courses. Furthermore, the manuscripts should provide sufficient motivation, examples and applications. This clearly distinguishes Lecture Notes from journal articles or technical reports which normally are very concise. Articles intended for a journal but too long to be accepted by most journals, usually do not have this "lecture notes" character. For similar reasons it is unusual for doctoral theses to be accepted for the Lecture Notes series, though habilitation theses may be appropriate.

2. Besides monographs, multi-author manuscripts resulting from SUMMER SCHOOLS or similar INTENSIVE COURSES are welcome, provided their objective was held to present an active mathematical topic to an audience at the beginning or intermediate graduate level (a list of participants should be provided).

   The resulting manuscript should not be just a collection of course notes, but should require advance planning and coordination among the main lecturers. The subject matter should dictate the structure of the book. This structure should be motivated and explained in a scientific introduction, and the notation, references, index and formulation of results should be, if possible, unified by the editors. Each contribution should have an abstract and an introduction referring to the other contributions. In other words, more preparatory work must go into a multi-authored volume than simply assembling a disparate collection of papers, communicated at the event.

3. Manuscripts should be submitted either online at www.editorialmanager.com/lnm to Springer's mathematics editorial in Heidelberg, or electronically to one of the series editors. Authors should be aware that incomplete or insufficiently close-to-final manuscripts almost always result in longer refereeing times and nevertheless unclear referees' recommendations, making further refereeing of a final draft necessary. The strict minimum amount of material that will be considered should include a detailed outline describing the planned contents of each chapter, a bibliography and several sample chapters. Parallel submission of a manuscript to another publisher while under consideration for LNM is not acceptable and can lead to rejection.

4. In general, **monographs** will be sent out to at least 2 external referees for evaluation.

   A final decision to publish can be made only on the basis of the complete manuscript, however a refereeing process leading to a preliminary decision can be based on a pre-final or incomplete manuscript.

   Volume Editors of **multi-author works** are expected to arrange for the refereeing, to the usual scientific standards, of the individual contributions. If the resulting reports can be forwarded to the LNM Editorial Board, this is very helpful. If no reports are forwarded or if other questions remain unclear in respect of homogeneity etc, the series editors may wish to consult external referees for an overall evaluation of the volume.

5. Manuscripts should in general be submitted in English. Final manuscripts should contain at least 100 pages of mathematical text and should always include

   – a table of contents;
   – an informative introduction, with adequate motivation and perhaps some historical remarks: it should be accessible to a reader not intimately familiar with the topic treated;
   – a subject index: as a rule this is genuinely helpful for the reader.
   – For evaluation purposes, manuscripts should be submitted as pdf files.

6. Careful preparation of the manuscripts will help keep production time short besides ensuring satisfactory appearance of the finished book in print and online. After acceptance of the manuscript authors will be asked to prepare the final LaTeX source files (see LaTeX templates online: https://www.springer.com/gb/authors-editors/book-authors-editors/manuscriptpreparation/5636) plus the corresponding pdf- or zipped ps-file. The LaTeX source files are essential for producing the full-text online version of the book, see http://link.springer.com/bookseries/304 for the existing online volumes of LNM). The technical production of a Lecture Notes volume takes approximately 12 weeks. Additional instructions, if necessary, are available on request from lnm@springer.com.

7. Authors receive a total of 30 free copies of their volume and free access to their book on SpringerLink, but no royalties. They are entitled to a discount of 33.3 % on the price of Springer books purchased for their personal use, if ordering directly from Springer.

8. Commitment to publish is made by a *Publishing Agreement*; contributing authors of multiauthor books are requested to sign a *Consent to Publish form*. Springer-Verlag registers the copyright for each volume. Authors are free to reuse material contained in their LNM volumes in later publications: a brief written (or e-mail) request for formal permission is sufficient.

Addresses:
Professor Jean-Michel Morel, CMLA, École Normale Supérieure de Cachan, France
E-mail: moreljeanmichel@gmail.com

Professor Bernard Teissier, Equipe Géométrie et Dynamique,
Institut de Mathématiques de Jussieu – Paris Rive Gauche, Paris, France
E-mail: bernard.teissier@imj-prg.fr

Springer: Ute McCrory, Mathematics, Heidelberg, Germany,
E-mail: lnm@springer.com

Printed in the United States
by Baker & Taylor Publisher Services